UPTAKE OF IONS
BY PLANT ROOTS

Uptake of ions by plant roots

D.J.F. BOWLING

Department of Botany,
The University of Aberdeen

LONDON

CHAPMAN AND HALL

A Halsted Press book
John Wiley & Sons, Inc., New York

First published 1976
by Chapman and Hall Ltd
11 *New Fetter Lane, London EC4P 4EE*

© 1976 *D. J. F. Bowling*

Typeset by Preface Ltd, Salisbury, Wilts
Printed in Great Britain by
Fletcher & Son Ltd,
Norwich

ISBN 0 412 12000 3

Distributed in the U.S.A. by Halsted Press,
a Division of John Wiley & Sons, Inc., New York

Library of Congress Cataloging in Publication Data

Bowling, Dudley James Francis, 1937—
    Uptake of ions by plant roots.

    Includes bibliographical references.
    1. Roots (Botany) — Physiology. 2. Plant
translocation. 3. Absorption (physiology) I. Title.
QK720.B67   1976   581.1   75-19244
ISBN 0-470-09285-8

Contents

Preface

I have approached the uptake of salts by the root by treating it is as a catena or chain with four links. These are: movement of salts in the soil to the root surface; uptake into the root; transport across the root and movement in the xylem to the shoot. The steps in the catena are dealt with in chronological order as separate chapters interspersed with chapters on the more theoretical aspects of salt uptake. In the final chapter I have attempted to sum up the situation prevailing in one plant under a given set of conditions. There has been a lot of work done on the individual steps in the salt uptake chain and I feel that an approach which brings together this knowledge and looks upon salt uptake as a whole plant process is certain to be of value in the future. Just as a chain is as strong as its weakest link the overall rate of salt uptake by a plant will depend on the rate of working of one of the component processes in the catena. Identification of the limiting step and knowledge of its characteristics would enable predictions about the salt relations of the whole plant to be made.

It has been said that plant cells are not squid axons or

blood cells. Likewise, it must be emphasized that root cells are neither algal cells nor green leaf cells. I have therefore tried to resist the temptation to extrapolate conclusions from the large celled algae, whose salt relations are comparatively well known, to the situation in the root. It has been my aim to outline the direct knowledge we have about ion uptake by roots and in some areas this turns out to be surprisingly sparse. Thus, the reader will not find a chapter specifically on membranes. This was omitted because I did not feel that we know enough about the relationship between membrane structure and salt transport in root cells. Whilst endeavouring to provide a balanced account of our knowledge in this field within the scope I have set myself, some undue bias and personal foibles will inevitably be apparent. Naturally I have tried to be as accurate in reporting as I possibly can and I apologise to anyone whose work I may have misinterpreted or misrepresented. I have attempted to underline what appear to be glaring gaps in our knowledge and tried to suggest possible lines for further research in some areas. If there are those who are encouraged to try to fill in some of these gaps after reading this book then it will have succeeded in its purpose.

Just as salt uptake by the plant is a catena so is the making of a book. I am conscious that I am but one link in a chain and so gratefully acknowledge the help and influence of many people in bringing this work into being. I am particularly indebted to my mentor Professor Paul E. Weatherley for introducing me to this subject and for subsequent encouragement along the way. Dr. Frank Cusick, Dr. Ian Alexander, Dr. James Hart and Dr. Alan Macklon read parts of the manuscript and made many helpful comments and suggestions. Dr. Cusick drew Fig. 1.1 and Dr. Alexander provided the photographs in Plates 1 and 3. Dr. Richard Johnson and David John prepared and photographed the material for Plate 2 and Dr. James Dunlop provided the photographs in Plates 4 and 5. Lorna Forbes and Susan Fraser drew the figures, Edward Middleton mounted some of the photographs and Hella Murray typed the legends. Finally I am greatly indebted to my wife for spending long hours typing the many drafts of the manuscript.

D.J.F.B.

Aberdeen
February, 1975

rata

e 26 line 2
mg kg^{-1} *read* m mol kg^{-1}
for 548 Ca *read* 54.8 Ca

Publisher's note

The units used in this book are SI (Système Internationale) or derived SI wherever possible. An exception is the use of mEq as a unit for cation exchange capacity. The litre was redefined, in 1964, to equal one cubic decimetre (dm^3) and this unit is generally used as a unit of volume in this text.

1 The root as an absorbing organ

Morphology

Sixteen or seventeen elements are required by the higher plant for growth and reproduction. Of these only three, carbon, hydrogen and oxygen are taken in through the aerial parts of the plant. The rest are taken in from the soil by the root. Certain elements, the so-called macronutrients consisting of calcium, potassium, magnesium, nitrogen, sulphur and phosphorus are required in relatively large quantities. Others including iron, boron, manganese, copper, zinc, molybdenum and chlorine are required in small amounts and are termed micronutrients. Sodium is also essential for some species growing in saline habitats. Some of the required elements, especially the micronutrients, are often present in the soil in only very small amounts. The fact that roots can find and absorb these elements, often accumulating them to high concentrations in the tissues, illustrates the efficiency of the root as an absorbing organ. Epstein (1972b) echoes the sentiments of a number of investigators before him when he describes the root as mining the soil for minerals.

1

Roots fulfil a number of roles besides absorption and their morphology varies widely depending on the part they play in the life of the plant. The most basic type of root system is the tap root which consists of a main axis and finer branches. The main root is often called the primary root and its branches, secondary roots. Secondary roots may in turn bear branches termed tertiary roots. This type of branching can be seen in the root of Sitka spruce shown in Plate 1. Quaternary roots have also been observed (Dittmer, 1948). The main axis of the tap root may penetrate to great depths in the soil and provide the plant with a firm anchorage. In some species the primary axis of the tap root is greatly enlarged and fulfils a storage function. Absorption of water and mineral ions is largely carried out by the secondary and tertiary roots which are produced acropetally just behind the growing point.

Another important type of root system is the fibrous root which, unlike the tap root, has no clearly defined main axis. Fibrous root systems are common in monocotyledonous plants such as the grasses which include many of the common cereals. In barley, for example, the root of the seedling, which develops from the embryonic root, is a mass of thin axes which are copiously branched. These are called seminal roots. In older plants, roots also develop from the base of the stem. These are often of larger diameter than the seminal roots and possess fewer branches. They are termed nodal roots. The branches which develop on seminal and nodal roots are called laterals.

On a dry weight basis the root usually makes up less than 50 per cent of the plant (Curtis and Clark, 1950) but its surface area is invariably much higher than that of the shoot. Dittmer (1937) made a careful study of the roots of a single rye plant and calculated that the surface area of the subterranean parts was 130 times that of the top. To obtain such high surface areas, considerable lengths of root are developed. Young elm trees 4—6 years old were found by

Dittmer (1948) to have root systems of about 2 km in length. Roots with an absorbing function tend to be near the surface of the soil and so the lateral spread of the root system is usually much greater than that of the aerial parts of the plant.

Physiological anatomy

If we look at a cross-section of a young root about 1 cm from its tip we find a relatively simple anatomy compared to that of the shoot. This may be related to the comparative uniformity of the subterranean habitat. Moving from the outside to the centre of the root we find four tissues, the epidermis, the cortex, the vascular system and sometimes the pith (Fig. 1.1).

The epidermis which forms the external surface of the root consists of a single layer of longitudinally elongated cells. A characteristic of the epidermis is the production of root hairs by the extension of the outward facing wall (see Plate 1). Root hairs may be absent under some conditions, for example, they do not develop in some roots when they are grown in water culture but in normal conditions in the soil they are usually present. Dittmer (1949) examined 37 species and found root hairs present in all of them. They develop in mature cells behind the growing point and may persist for several centimetres along the root but they usually die off in the older parts. The number of hairs developed can be quite high, Haberlandt (1914) quotes a density of 425 root hairs per mm^2 in maize and 232 in pea. Drew and Nye (1969) found an average of 98 root hairs per mm length of root in rye grass.

The cortex is the largest root tissue by volume and consists of concentric layers of large cells with numerous air spaces between them. The cell walls are usually unthickened but in some monocotyledons lignification occurs. The cortex is very

3

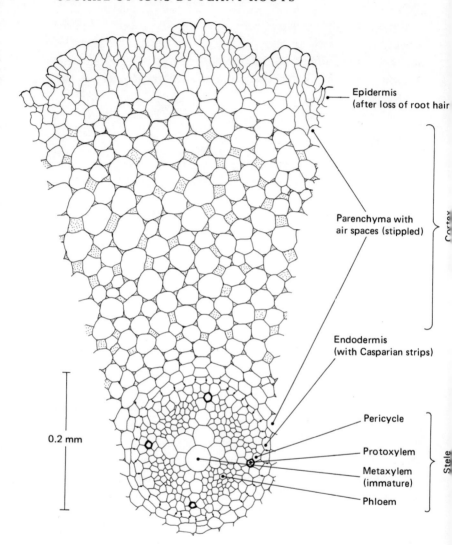

Epidermis
(after loss of root hair

Parenchyma with
air spaces (stippled)

Cortex

Endodermis
(with Casparian strips)

Pericycle

Protoxylem

Metaxylem
(immature)

Phloem

Stele

0.2 mm

Fig. 1.1

Transverse section of a root of *Ranunculus repens* taken approximately 1 cm from the apex. Drawn by Dr Frank Cusick.

variable in size. In primary roots it may be up to 15 cells deep but in tertiary roots there may be as few as five cells between the epidermis and the vascular tissue. Protoplasmic connections, the plasmodesmata, may be observed between the cells but precise data on their frequency is lacking. Tyree (1970) quotes a figure of 1.5×10^8 cm^{-2} in mature cortical cells of onion, a value computed from the data of several workers.

The vascular cylinder is surrounded by a compact layer of cells which forms a distinct boundary separating the vascular tissue from the cortex. This is the endodermis which is invariably present in roots and which anatomists regard as being the innermost layer of the cortex. Its chief characteristic is the presence of a strip or ribbon-like structure on the radial walls of each cell (Figs. 1.1 and 1.2). This is called the Casparian band after its discoverer R. Caspary. The chemical composition of the Casparian band is not very clear but it appears to be made up of a mixture of cellulose and suberin (Van Fleet, 1961).

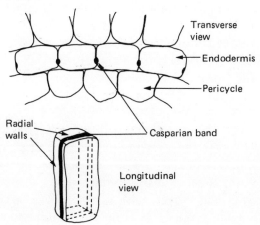

Fig. 1.2
Diagram to show the location of the Casparian band in the endodermis.

Circumstantial evidence strongly suggests that the endodermis with its Casparian band is a barrier to water movement in the cell walls between the cortex and the stele but good experimental evidence for this is lacking. The suberin in the Casparian band is presumed to block the capillaries between the microfibrils of the radial walls of the endodermis. Water and salts on their way to the stele must therefore pass across the membranes of the cytoplasm in the endodermis and this provides the root with a means of controlling salt and water movement.

If the endodermis plays an active role in transporting ions and water by, for example, secretion, it might be expected that the electron microscope would reveal a structural basis in the cytoplasm for such activity. However, there is no evidence that the endodermal cells have any peculiarities in their cytoplasmic structure. Bonnett (1968) investigated the endodermis of *Convolvulus arvensis* with the electron microscope but found no evidence in its fine structure for a secretory role. He did, however, observe plasmodesmata connecting the endodermal cells with all the neighbouring cells. Helder and Boerma 1969) in an electron microscopical study of young barley roots calculated that there are approximately 20 000 plasmodesmata in each tangential wall of the endodermis. This means that there are 5×10^8 plasmodesmata per cm^2 a figure which has been confirmed by Clarkson, Robards and Sanderson (1971). Thus, although there is little evidence that the endodermis takes an active role in the transport of salts the presence of large numbers of plasmodesmata indicate that it is well adapted for a more passive regulatory role in salt transfer to the stele.

The young endodermis with a fully developed Casparian band has been termed the primary endodermis. In older roots, particularly those that overwinter, the radial and inner tangential walls of the endodermis often develop secondary deposits of suberin and tertiary deposits of cellulose and sometimes also of phenolic and quinoidal substances. In these cells the transport of substances through the walls presumably ceases but often a few cells remain without these secondary and tertiary deposits and probably continue to serve a function in transport. These are called passage cells which, although lacking secondary and tertiary thickening, possess the Casparian band and are assumed to behave exactly like the primary endodermis. In fact the primary endodermis can be looked upon as being wholly made up of passage cells.

Clarkson, Robards and Sanderson (1971) observed the development of tertiary endodermis in the basal zones of the barley root approximately 45 cm from the apex. They found that up to 40 per cent of the phosphate and water transported to the shoot passed through basal parts of the root with tertiary endodermis. However, passage cells were very infrequent, there being about one passage cell to every thousand endodermal cells. Calculation of the flux rates assuming that all the ions and water moved by way of the passage cells gave unrealistically high values and so it appeared that most of the transport was occurring across the thickened cells. The electron microscope revealed large numbers of plasmodesmata permeating the thickened areas of the endodermis. Except for some constriction at their ends due to the thickening, they appeared to be similar to plasmodesmata in unthickened cells. Assuming them to be the main pathway for salts and water, Clarkson *et al.* calculated a phosphate flux of $1.8-9.7 \times 10^{-9}$ mol s^{-1} cm^{-2} and a water flux of $1.20-4.78 \times 10^{-4}$ cm^3 s^{-1} cm^{-2}. It would be very interesting to know what proportion of the salts and water transferred across the tangential walls of primary endodermal cells is carried by the plasmodesmata.

The stele, the central cylinder of the root is surrounded by a single or multiseriate layer, the pericycle, which lies immediately inside the endodermis. It consists of thin walled cells

7

and is usually the place of origin of the lateral roots. The conducting tissues lie inside the pericycle. Xylem usually alternates with phloem and may also occupy the centre of the root. If there is no xylem developed in the centre, it is occupied by pith, a tissue usually composed of large cells with thin walls. The first xylem to mature, the protoxylem, occurs on the outside of the xylem strand whilst the later maturing metaxylem, which is characterized by large vessels, is found towards the inside. Mature vessels are without cell contents and their walls are usually thickened with a deposit of lignin which increases the strength of the cells enabling them to withstand the tensions set up during transpiration. Stelar structure is illustrated in Plate 2.

Xylem vessels are usually surrounded by less differentiated xylem cells, the xylem parenchyma which resemble the cells of the cortex but are often smaller. Läuchli, Spurr and Epstein (1971) made an ultrastructural study of these cells in maize roots and observed the presence of numerous membrane systems. They suggested that the xylem parenchyma may secrete ions into the adjacent xylem vessels. They drew a parallel between these cells and transfer cells, a term given to cells which appear to be specialized in the transfer of solutes because of the presence of folded membranes affording a large surface area for transport. Transfer cells were first recognized by Gunning and Pate (1969) who have identified them in various parts of the plant but they have so far not been observed in normal roots. They do however, occur in some roots specialized for nitrogen fixation which possess nodules containing nitrogen fixing micro-organisms. Here they appear to be involved in transfer of the products of nitrogen fixation to the aerial parts of the plant. However, in a more recent study, Läuchli, Kramer, Pitman and Lüttge (1974), investigated the structure of the xylem parenchyma cells of barley with the electron microscope and although they observed numerous plasmodesmatal connections with the adjacent living cells no structures were found which would suggest that these cells behave like transfer cells. So the role of the xylem parenchyma in salt transport is still an open question.

Bundles of phloem alternate with the xylem in the young root. The phloem is a complex tissue which consists of a number of cell types including the sieve tubes that transport carbohydrates and other organic compounds from the leaves. In roots there does not appear to be any upward movement of substances in the phloem and water and salts are transported from the root to the leaves almost exclusively in the xylem. However the phloem is known to transport ions exported from the leaves back down to the root. In perennial dicotyledons, secondary xylem and phloem are often developed at the end of the first growing season but as salt uptake is largely confined to roots with primary vascular tissues, secondary thickening need not be considered here.

Development

The tip of the root is covered by the root cap, the function of which is to protect the growing point which lies immediately behind it. Growth is restricted to the root apex and is brought about by cell division in the meristematic zone followed by cell elongation further back from the apex. Analysis of root growth in *Phleum pratense* by Goodwin and Avers (1956) showed that the extending portion of the root was approximately 1.22 mm long, excluding the root cap. Therefore, although its length varies between species the growing zone forms a comparatively small proportion of the root.

Phloem begins to reach maturity before elongation has ceased but protoxylem matures only after elongation has been completed. There is some difference of opinion about the timing of endodermal development. According to Esau (1965) the Casparian band appears at the same time as, or a

Table 1.1

Barley root maturation based on data from Heimsch (1951) and Erickson and Sax (1956).

Tissue	Maturation-distance from apex (mm)
Protoxylem	0.25—0.75
Protophloem	0.4—8.5
Metaxylem	0.55—21.6
Casparian band	0.75

little later than, the time of protoxylem maturation. Fahn (1967) on the other hand considers that the Casparian band develops before the maturation of the protoxylem. Root hairs only develop after elongation of the epidermal cells has ceased. In contrast to most of the other cells, metaxylem vessels may still be maturing several centimetres from the apex (Table 1.1). Anderson and House (1967) have reported the presence of xylem elements with cytoplasm up to 10 cm from the tip in maize roots. Higinbotham, Davis, Mertz and Shumway (1973) confirmed this observation and published photographs of metaxylem vessels at 9.5—10.0 cm from the apex of maize roots with cytoplasm and nuclei. The rate of maturation of metaxylem appears to be variable as the data for barley in Table 1.1 indicate, and Läuchli *et al.* (1974) observed that in their barley material metaxylem vessels had no living contents at only 1 cm behind the apex.

The root in the soil

The soil can be looked upon as a three component system, the soil particles and decaying organic matter form the solid phase and the space between is occupied by the soil solution and gaseous phase.

The nature of the solid phase varies widely depending on the origin of the soil but the particles often contain colloidal miscelles which present a large surface area and are capable of holding large amounts of ions. The ions bound on the soil colloids are usually exchangeable and this phase is called the exchange complex. Ions from the exchange complex dissolve in the soil solution until an equilibrium between the solid phase and the soil solution is set up. Both cations and anions are held in the exchange complex which is therefore amphoteric, but some anions such as NO_3^-, SO_4^- and Cl^- are only weakly held and so may be present in the soil solution at relatively high concentrations. Phosphate, on the other hand is strongly held in the exchange complex and hence is usually present in the soil in relatively small amounts.

The composition of the soil solution varies widely depending on factors such as the mineral basis of the soil, the amount of organic matter present and pH. Table 1.2 shows

Table 1.2

Ionic composition of the soil water of John Innes no. 2 potting compost at field capacity. The soil water was obtained by centrifuging the soil for 5 minutes at 100 g. Data from Ansari and Bowling (1972).

Ion	Concentration ($mmol\ dm^{-3}$) ± standard error of mean
K^+	0.75 ± 0.10
Na^+	2.1 ± 0.1
Cl^-	4.2 ± 0.3
NO_3^-	2.0 ± 0.3
SO_4^{--}	13.9 ± 3.7
Mg^{++}	1.4 ± 0.1
Ca^{++}	7.1 ± 0.9
pH	5.9 ± 0.2

the composition of the soil solution obtained by centrifugation of John Innes potting compost (see also Table 2.3). This potting compost is a medium rich in ions and natural soils generally have much lower mineral contents in their soil solution. Phosphate and the micronutrients were not determined but they are generally present in the soil solution at much lower concentrations than those of the macronutrients. Phosphate for example, is usually present at $10^{-2}-10^{-3}$ mmol dm^{-3} (Fried and Broeshart, 1967).

The gaseous phase of the soil is very important to the root as it must contain enough oxygen to support respiration. The soil usually contains large numbers of organisms such as fungi, bacteria and various small animals, which compete along with the roots for the oxygen in the soil atmosphere. Root growth is restricted if the oxygen content of the surrounding medium drops below 5 per cent (Erickson, 1946). Micro-organisms also compete with the root for mineral nutrients and in some circumstances they can have a decisive effect on plant nutrition. Furthermore, the root in natural conditions in the soil is frequently covered with a mat of fungal mycorrhiza. Dittmer (1949) in a study of the roots of 37 species found a mantle of ectotrophic fungal hyphae on practically all the plants he examined. Mycorrhiza are known to play an important part in the mineral nutrition of plants particularly their phosphorus nutrition. This will be discussed more fully in Chapter 3. Plates 1 and 3 show mycorrhiza on roots of Sitka spruce.

The root and the soil surrounding it form a highly complex and dynamic system. The root grows between the soil particles, continually occupying new areas of the soil. Goodwin and Avers (1956) found that the root of *Phleum* extended approximately a millimetre every three hours. The root cap sloughs off cells as it goes and produces a mucilage which may act as a lubricant to aid its passage. This mucilage, termed mucigel by Jenny and Grossenbacher (1963), may fill the space between the cell wall and the soil particles. Electron micrographs have shown that there is intimate contact between the mucigel and the soil particles. Behind the elongating region, root hairs begin to develop and may grow out to a length of up to 1500 μm (Dittmer, 1949) approximately doubling the (already vast) surface area of the root (Dittmer, 1937; Drew and Nye, 1969).

Water culture

The soil-root association has proved very difficult to study because of its complexity. Most plant physiologists investigating the uptake of ions by roots have taken an easy way out and ignored the soil altogether. Instead they have made use of the water cuture technique developed in the middle of the nineteenth century by the classical plant physiologists Sachs, Knop and Nobbe. This involves the use of a solution containing all the essential elements. The method provides the root with a medium similar in composition to the soil water without the complication of the soil particles. A large number of different culture solution recipes have been used over the

Table 1.3

Composition of a balanced nutrient solution for water culture.

Macronutrients	(mmol dm^{-3})	Micronutrients	(mmol dm^{-3})
KNO_3	6	B	5×10^{-2}
$Ca(NO_3)_2$	4	Mn	1×10^{-2}
$MgSO_4$	2	Zn	1×10^{-2}
KH_2PO_4	1	Cu	0.2×10^{-3}
Fe EDTA	0.2	Mo	0.1×10^{-3}
(Iron chelate)			

N.B. A trace of Cl, an element essential for growth is introduced with Fe EDTA as $FeCl_3$ is used together with ethylenediaminetetra-acetic acid (EDTA) to make the iron chelate.

13

years and Table 1.3 gives the composition of a solution used successfully by the author.

The culture solution technique enables the composition, pH and aeration of the root medium to be closely controlled. Furthermore, the rate of disappearance of an ion from the solution can be used as a measure of uptake by the root. There are however, some disadvantages with the technique. The behaviour of the root in culture is likely to be different from that in the soil because its morphology is different. For example, root hairs commonly present in soil-grown plants are frequently absent on roots grown in water culture. The microflora of the root surface is also likely to be different. Mycologists have often criticized plant physiologists for studying roots in water culture which are without the mantle of mycorrhizal fungus which usually develops in soil grown roots. Therefore extrapolation of results and conclusions obtained with water culture-grown roots to roots growing in the soil has to be made with caution. Having said that, however, it must be recognized that the water culture technique has proved to be a very useful tool in enabling us to understand some of the processes which occur during salt uptake. A number of different water culture systems have been devised to try to overcome some of the disadvantages mentioned above, and for details of these the reader is referred to the work by Hewitt (1966).

2 The transfer of ions across the soil–root interface

The cell wall

Ions moving from the soil to the root have to pass through the cell wall before they can be taken up into the cell. A knowledge of the structure of the cell wall and its ability to allow the passage of ions is therefore very important. Its structure has been likened to that of reinforced concrete, with cellulose as the framework material, embedded in a matrix of pectins and hemicelluloses, and in many secondary tissues, lignin. Waxes, fats, proteins and various pigments may also be present. The proportions of the various constituents show wide differences between species and indeed between different organs of the same species. There appears to be little information about the cell walls of roots but primary root cell walls are likely to be very similar in constitution to the walls of the *Avena* coleoptile given in Table 2.1.

Cellulose is the most important constituent of the cell wall and consists of β-1,4 linked glucose units forming what are known as glucan chains. Up to 100 glucan chains group together to form elementary fibrils. Approximately 20 elementary fibrils are organized into a microfibril about 8–10 nm in

15

Table 2.1

Constituents of the cell wall of *Avena* coleoptile. Data of Bishop, Bayley and Setterfield (1958)

Component	%
Cellulose	24.8
Hemicellulose	50.7
Pectin	0.3
Lignin	—
Waxes and Fats	4.2
Protein	9.5

diameter which is clearly visible under the electron microscope. The microfibrils are interwoven in the primary cell wall and arranged in parallel in the secondary wall. They are relatively widely spaced and one microfibril is usually separated from the next by about four times its own diameter. Water occupies a large proportion of the space between the microfibrils, making the wall highly permeable to water soluble substances.

From the point of view of ion permeability the pectins are very important. They consist of polymers of D-galacturonic acid, L-arabinose D-galactose and L-rhamnose. The carboxyl groups of the D-galacturonic acid residues in the pectin molecules are normally dissociated and so leave the wall with a negative charge. This charge gives the wall tremendous cation binding capacity.

Donnan phase

The fixed negative charge on the cell wall attracts cations such as Ca^{++} and K^+. This region of the wall containing the negative sites is called the Donnan phase. The planar

Cell
wall

Electrical
double layer

Fig. 2.1

Diagrammatic representation of the arrangement of charges on the surface of the cell wall.

arrangement of the fixed negative charges attracts a parallel file of positive charges and an electrical double layer is formed (Fig. 2.1).

Let us consider a situation where the diffusing cations come into equilibrium with the fixed negative charges on the wall. Consider the ions of a mobile salt $A^+ B^-$. A_1^+ will move to balance the charge on a fixed anion X^-, leaving its accompanying anion B_1^-. (Fig, 2.2). Another cation A_2^+, from further away, will move into balance the charge on B_1^-, but will be restricted from doing so by B_2^-. At equilibrium an electrical potential difference will be set up at the wall surface with the wall negative compared to the surrounding solution. This PD is termed a Donnan potential. At equilibrium the Donnan potential (ΔE) can be expressed as follows:

$$\Delta E = RT \ln\frac{aA_1^+}{aA_2^+} \qquad (2.1)$$

where R = the gas constant, T = absolute temperature and a = the chemical activity of the ion A.

17

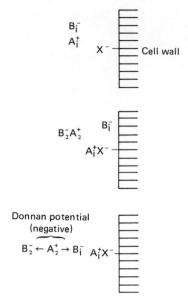

Fig. 2.2

Origin of the Donnan potential at the cell wall.

Similarly

$$\Delta E = RT \ln \frac{a[B_2^-]}{a[B_1^-]} \tag{2.2}$$

$$\therefore \frac{a[A_1^+]}{a[A_2]} = \frac{a[B_2^-]}{a[B_1^-]} . \tag{2.3}$$

If the activity coefficients of A^+ and B^- are close to one then

$$[A_1^+] \times [B_1^-] = [A_2^+] \times [B_2^-]. \tag{2.4}$$

Therefore, although

$$[A_1^+] < [B_1^-] \text{ and } [B_2^-] < [A_2^+]$$

their products are equal.

Donnan potentials should develop on both sides of the cell wall but as they are in opposite directions they should cancel each other out making the net PD across the wall zero. The Donnan potential should favour diffusion of cations through the wall but tend to repel anions, thus restricting their diffusion. This might be one of the reasons why diffusion coefficients of some molecules are up to 100 times lower in the cell wall than in aqueous solution, even though water forms the largest proportion of the cell wall by weight (Nobel, 1970).

The concentration of fixed negative charges in the Donnan phase can be estimated using radioisotopes. Briggs, Hope and Pitman (1958) pretreated discs of beet tissue with solutions of rubidium iodide to remove all other mobile ions from the wall. The amount of exchangeable I and Rb was then determined by measuring the uptake of $^{131}I^-$ and $^{86}Rb^+$ at different external concentrations. The excess of cations over anions was used to estimate the amount of non-diffusable anions in the Donnan phase. This was found to be approximately $10-13$ mmol kg^{-1}. Using an ingenious technique in which the extra exchangeable Rb^+ was determined in discs with the Donnan phase occupied by Ca^{++}, they estimated the apparent volume of the Donnan phase. This amounted to 2.1 per cent of the tissue, thus the concentration of non-diffusable ions in the Donnan phase was estimated to be 560 mmol dm^{-3}. They called this volume the Donnan Free Space (DFS).

Briggs et al. considered that the number of immobile anions present in beet tissue was too great to be contained in the walls and concluded that the DFS also extended to the cytoplasm. Dainty and Hope (1959) however, have shown fairly conclusively that the DFS as measured by isotope exchange is restricted to the cell wall. They studied ion exchange in the cells of the alga Chara australis. After periods of equilibration of up to seven days in artificial pond water

19

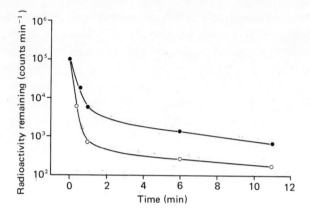

Fig. 2.3

Radioactivity remaining in cells and cell walls of
Chara australis plotted on a logarithmic scale against
time. (O ——— O); wall; (● ——— ●); cells. Plotted
from data of Dainty and Hope (1959).

(APW), the cells were placed in artificial pond water labelled
with $^{22}Na^+$, $^{45}Ca^{++}$ or $^{131}I^-$. After a time the cells were
removed, blotted and the radioactivity eluted into aliquots of
inactive APW (Fig. 2.3). They were able to compare the
behaviour of whole cells with that of the cell walls by
scraping out the cell contents. They found that cation
exchange from the cells and cell wall was complex as can be
seen from Fig. 2.3. The elution pattern of $^{131}I^-$ was simpler.
Assuming that the space occupied by iodine was in the
aqueous phase of the wall, and that at equilibrium the
concentration of iodine in the wall space was equal to that in
the external solution, they could calculate the volume of the
Iodine Free Space (IFS). This was found to be
46 per cent of the total wall water. From their cation
exchange measurements they calculated that the average
concentration of non-diffusable anions was 0.32 μmol dm^{-3}
wall water. However 46 per cent of the total wall water is

WFS and therefore the concentration of the fixed negative charges in the DFS is $0.32/0.54 = 0.6 \ \mu\text{mol dm}^{-3}$. This value is very similar to that obtained by Briggs *et al.*

The two components of the free space, the water free space and the Donnan free space together form what has been called the Apparent Free Space (AFS).

$$\text{Thus AFS} = \text{WFS} + \text{DFS} \qquad (2.5)$$

The word *apparent* is important because the term AFS concerns experimental estimates and emphasizes the approximate nature of such estimates. The value of the WFS has been found to depend on the ion used to estimate it and this may also apply to DFS because it is possible that the negative sites on the wall may have different affinities for different cations. The magnitude of AFS for anions is smaller than that for cations as there is of course no DFS component for anions. AFS measurements for roots are usually given as a percentage of the total volume of the tissue. Methods for measurement of AFS and the importance of the concept in our understanding of ion transfer across the root are discussed in Chapter 7.

Cation exchange capacities of roots and soil

Plant physiologists and biophysicists have made measurements of DFS only recently but soil scientists have known since the early part of this century that plant tissues possess cation exchange properties (Devaux, 1916). Soil scientists measure this characteristic as Cation Exchange Capacity (CEC). Whilst all the main organs of the plant show a cation exchange capacity, the soil scientist has naturally paid most attention to CEC in roots.

Methods of measuring CEC aim initially to remove all

adsorbed ions from the root and to replace them by H^+. This is usually done by immersing the roots in strong hydrochloric acid. One method for determining the number of negative sites on the root surface is to replace the H^+ on 'H^+ roots' by Ca^{++} and this is again displaced by further acid washing and determined by flame photometry or counting if labelled Ca^{++} has been used. A more direct method is to place the 'H^+ roots' in a solution of a neutral salt such as KCl and the amount of alkali required to restore the system to neutrality gives a value for CEC (Drake, Vengris and Colby, 1951; Crooke, 1964). CEC is usually expressed as mEq/100 g dry matter.

Each plant species appears to have its own characteristic CEC level which is independent of location (Crooke and Knight, 1971) and monocotyledons tend to have a lower CEC than dicotyledons as illustrated by the data of Crooke (1964) given in Table 2.2.

It is reasonably certain the CEC and DFS are contiguous. CEC appears to be restricted to the cell wall as killing the root with ether has no effect on the CEC values obtained (Williams and Coleman, 1950). Knight, Crooke and Inkson

Table 2.2

Cation exchange capacities of the roots of various species of monocotyledons and dicotyledons. Data of Crooke (1964). Cation exchange capacity (C.E.C.) is expressed as mEq/100 g dry matter.

Monocotyledons		Dicotyledons	
Species	C.E.C.	Species	C.E.C.
Arundinaria sp.	5.3	Viola riviniana	20.2
Festuca ovina	13.1	Ilex pyramidalis	24.0
Iris sp.	21.9	Helleborus sp.	37.7
Knipholia sp.	23.6	Trollius sp.	53.2

(1961) determined CEC for about 80 species and found a close correlation between CEC and uronic acid content.

Does CEC have any role to play in the process of ion uptake or is it just an unimportant consequence of cell wall structure? Drake *et al.* (1951), after examining over 30 species, suggested that the difference in ability of plants to take up cations from the soil was controlled by the cation exchange capacity of the root. Crooke and Knight (1962) made an evaluation of published data on the mineral composition of plants and CEC and showed that CEC of roots is positively correlated with the content of the tops of:

(a) total cations
(b) ash
(c) excess base
(d) total trace elements

Rao, Krishnamurthy and Rao (1967) determined CEC for 11 varieties of sugar cane. They found a high correlation (0.87) between CEC of sett roots and the mean yield of cane (Fig. 2.4). Crooke and Knight (1971) have also reported similar results for a large number of leek varieties.

Such impressive correlations however do not enable us to distinguish between cause and effect. Results such as those described above have to be weighed against evidence for the factors which control ion uptake into the cytoplasm and vacuole of the cell. When this is done it is clear that metabolically controlled transport across the cell membrane is the most important factor regulating salt uptake by roots (see Chapters 3 and 4). This type of transport is markedly affected by environmental factors such as temperature and is highly selective, whereas physical adsorption on to the cell wall is relatively unaffected by temperature (Williams and Coleman, 1950) and shows little selectivity (Broeshart, 1962). Under natural conditions divalent ions such as Ca^{++} usually predominate in the Donnan free space, yet potassium

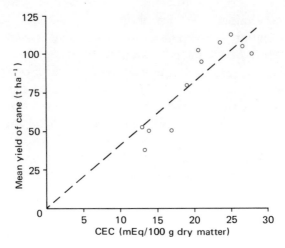

Fig. 2.4

Relationship between cation exchange capacity
(CEC) of sett roots of 11 varieties of sugar cane
and yield of cane. Each point represents the mean
values for each variety. Data from Rao,
Krishnamurthy and Rao (1967).

is accumulated by cells at rates some 20–60 times greater
than those for Ca^{++}, indicating that there is no relationship
between the exchange capacity of the wall and subsequent
accumulation in the cytoplasm and vacuole. It appears,
therefore, that CEC and salt accumulation, rather than being
causally related are products of a common process, namely
the overall metabolic activity of the plant.

Soil-root transfer

It is generally agreed that ions reach the root surface from
the soil by way of the soil solution. The possibility of there
being a direct transfer of cations from negative sites on the
soil colloids to the Donnan free space of the root cell wall, as
postulated by Jenny and Overstreet (1939), has now been

discounted through lack of positive evidence. In fact some evidence against the contact exchange theory, as it is called, has been obtained by Olsen and Peech (1960). They measured uptake of Rb^+ and Ca^{++} by barley and mung bean roots from clay suspensions and cation exchange resins. They compared their results with uptake measurements from dialysates of the exchange resin and clay media. Despite the fact that the concentration of the ions on the clay and the exchange resins was between 53 and 6600 times greater than in the corresponding equilibrium solutions they found no difference in uptake from the two phases.

Ions can move to the root surface in the soil water in two ways, by mass flow of the soil solution or by diffusion (Barber, 1962; Drew, Vaidyanathan and Nye, 1967). The mass flow component can be estimated from the product of the soil solution concentration and the volume of water transpired by the plant. This, however, does not take into account mass flow of the soil solution which is not due to the water uptake by the root such as gravity induced movement of water down the soil profile. It should be noted that movement of ions by mass flow *up to* the root, but not *into* it is being considered here. The latter will be discussed in Chapter 7. If uptake by the root is greater than can be supplied by the bulk flow of soil solution then the concentration of the ion or ions concerned will be lowered at the root surface and the soil solution further away and diffusion will then become more important.

This raises the question of the relative importance of mass flow and diffusion in supplying the root with nutrients under field conditions. Barber, Walker and Vasey (1962) provided an answer to this question for the American corn crop by means of a simple calculation. They determined the concentration of several major nutrient ions in the soil solution of a number of soils as shown in Table 2.3. It was assumed that corn transpires 500 g of water for each gramme of dry

Table 2.3

Concentrations of some major ions in the soil water of a number of soils in Indiana and Georgia (means mmol dm^{-3}). Data of Barber, Walker and Vasey (1962).

Depth	Ca	Mg	K	P
0—15 cm	0.58	3.13	1.30	0.005
46—51 cm	0.10	0.72	0.49	0.002

weight it produces. At harvest, maize contains on average (in mg kg^{-1}) 548 Ca, 512 K, 74 Mg and 65 P. Assuming that all the ions were supplied by mass flow, they calculated the ionic concentrations in the soil solution that would be required simply by dividing the above figures by 500. This gave (in mmol dm^{-3}) 0.11 Ca, 1.0 K, 0.15 Mg and 0.13 P. Clearly mass flow could supply all the plants needs for Mg^{++} and Ca^{++} but not enough K^+ and P, bearing in mind that the roots of maize penetrate the soil to depths greater than 15 cm.

The importance of diffusion as well as mass flow suggested by the above calculation has been verified using quite a different approach. Making a valuable contribution to the study of this problem, Barber and his colleagues developed a method of growing roots in boxes of labelled soil. One side of the box could be removed enabling autoradiographs to be made of the soil-root association. Using ^{86}Rb as a tracer for potassium they were able to show that large gradients quickly develop around the root (Barber, 1962; Walker and Barber, 1962).

Using labelled molybdenum, Lavey and Barber (1964) found that gradients in Mo concentration developed around roots which depended on the concentration of Mo in the soil. With concentrations above 0.004 ppm, mass flow caused accumulation of Mo at the root surface whilst concentrations below

0.004 ppm resulted in diffusion gradients being set up. Lewis and Quirk (1967) have made similar autoradiographs of phosphorus distribution around the root. They observed the development of diffusion gradients as predicted by transpiration ratio calculations. Their pictures show that these gradients also occur around the root hairs and suggest that the root hairs may be very important in the absorption of ions such as phosphate which have low mobility in the soil because of adsorption on the soil colloids. Nye (1966) has calculated that root hairs may be advantageous by increasing the surface area of the root, thereby bringing more phosphate within the reach of an absorbing surface.

Bhat and Nye (1973) have improved the autoradiographic technique for phosphorus by using, instead of ^{32}P, the low energy emitter ^{33}P which affords greater resolution. They were able to measure the phosphorus gradient around *Brassica napus* roots using densitometry. Using a standard diffusion model they calculated the amount taken up by the root and found that it compared well with the actual uptake of P.

Where demand by the root for potassium and phosphate is high the amount of labile potassium and phosphate in the immediate vicinity of the root and within a short distance of the root hairs is drastically reduced. Drew and Nye (1969) calculated that available K^+ within the root hair cylinder of rye grass was reduced by between 53 and 99 per cent in a four day period. During this time K^+ from the root hair cylinder was responsible for less than 10 per cent of the total potassium uptake, therefore considerable amounts of K^+ must have been diffusing to the root surface from outside the root hair zone. Diffusion over such long distances is relatively slow and so it must be a limiting factor, regulating potassium uptake by roots under most field conditions. Drew and Nye concluded, that in experiments in which demand was high, diffusion appeared to limit uptake by 59—71 per cent of that which the roots would be expected to absorb from stirred solutions.

Passioura's equation

Mass flow and diffusion usually occur simultaneously and any theoretical treatment needs to take this into consideration. Passioura (1963) attempted a mathematical model to describe the uptake of ions from the soil solution. He considered that eight parameters needed to be taken into account in order to derive an expression for F, the total iron flux to the root surface.

These are:

M, the root uptake coefficient, a measure of plant nutrient demand which has dimensions of a velocity.

F, which is equal to MC_0 where C_0 is the ion concentration at the root surface.

V, the water flux into the root.

C, the initial soil concentration of the ion.

D, the diffusion coefficient of the ion in the soil solution.

r, the root radius.

t, time.

f, the function of Dt/r^2 which is usually very close to unity.

The total flux of the ion (F) over unit surface area of root is then given by

$$F = \frac{(C - C_0)Df}{r} + \frac{CV}{t} . \tag{2.6}$$

It can be seen that equation (2.6) incorporates both the diffusional and mass flow components. It applies to a root which is not growing, i.e. the sink for ions is constant. For a root which is advancing through the soil at velocity v Passioura proposed an integrated derivative of equation (2.6).

However, it can be argued that though the root is growing, the actual size of the sink is not changing, or changing only slowly as the ion uptake is largely restricted to the young roots and declines with root age. Therefore equation (2.6) seems likely to provide an approximate description of the movement of ions to growing roots. Originally equation (2.6) was restricted to ions which are not adsorbed on to the soil colloids, such as nitrates and halides, but Tinker (1969) considered that it should apply to all ions which equilibrate rapidly between the solid and liquid phases of the soil.

The diffusion coefficient D is a critical parameter in equation (2.6) but it is not easy to measure. Values of D for soil are less than the corresponding values in water because the soil solution occupies pores in the soil which have a varying tortuosity. D also varies with soil moisture level and this is presumably related to the water film thickness around the soil particles (Rowell, Martin and Nye, 1967). The charges in the soil surface also affect the movement of the counterions thus altering the apparent value of D. The gradient $C-C_0$ is also difficult to define and may be constantly changing. Despite these obvious difficulties Nye and Marriott (1968) have found that results obtained with Passioura's equation compare favourably with computer solutions of the correct differential equation so it does appear that it can be used to predict the approximate conditions existing around the root. Tinker (1969) describes its successful application to potassium uptake by leek roots. He found that the potassium concentration at the root surface predicted by the equation was always about half that of the initial soil concentration indicating that, under the conditions of his experiments, diffusion was of considerable importance.

Passioura's model is based on reasonable assumptions but it does not take into consideration important variables such as the influence of root hairs, competition between roots and different parts of the same root or the proportion of the root which is taking part in ion uptake or the effect of water flux on the uptake ability of the root. This highlights the extreme complexity of the soil-root association. Many of the variables involved are still not properly understood. Investigation of the transfer of ions from the soil to the root has been neglected by soil scientists and avoided altogether by plant physiologists. More attention is required to this field before we can hope to understand this important stage in ion uptake by the root.

3 Accumulation in the vacuole

Some early investigations

Much of the early knowledge of ion uptake by plant cells was due to the work of the American plant physiologist D. R. Hoagland. Studies by Hoagland and Davis (1929) on the large celled alga *Nitella* showed that it could accumulate all the major ions in the vacuole at higher concentrations than those in the surrounding pond water. Conductivity measurements on the vacuolar sap showed that this accumulation could not be explained in terms of Donnan equilibria. Hoagland and Davis also found that light produced a marked stimulatory effect on bromide accumulation by *Nitella*. They obtained similar results with the marine alga *Valonia* and it became clear to Hoagland that ion uptake was not brought about merely by physical processes, as had been previously believed, but also involved the metabolism of the cell. Further evidence for the identification of salt uptake with cell metabolism was furnished by F. C. Steward and his co-workers on potato and artichoke tuber tissue (Steward, 1932, 1933; Steward and Berry, 1934; Steward, Berry and Broyer, 1936). Studies on the relationship between metabolism and

31

ion uptake by wheat roots were carried out at about the same time by Lundegårdh and Burström (1933, 1935) in Sweden.

The first evidence that roots could accumulate ions in the vacuole against a concentration gradient was reported by Hoagland and Broyer (1936) in an important paper. They used barley roots excised from plants grown in aerated water culture. They carefully controlled the volume of solution given to the plants and produced roots which were partially depleted of their salt content. Roots treated in this way proved to have the capacity to accumulate large amounts of K, Br and NO_3 during a ten hour period. Vacuolar sap was obtained by freezing and thawing the tissue and expressing the sap mechanically. The solution they obtained was obviously contaminated by sap from the cytoplasm but as the cytoplasm occupies a relatively small volume in the cell compared with the vacuole they concluded that the solution was a close approximation to the vacuolar sap. They found that the cells could accumulate K, NO_3 and halides to concentrations several times that of the culture solution (Table 3.1). Conductivity measurements indicated that the ions were in free diffusion in the vacuolar sap showing that the high accumulation ratios were not due to Donnan effects. This finding was very important as it virtually proved beyond doubt that metabolic processes are involved in ion accumulation by the root.

Table 3.1

Accumulation of potassium, nitrate and bromide by excised barley roots over a 10h period at $24°C$. From Hoagland and Broyer (1936).

	Potassium	Nitrate	Bromide
Concentration in culture solution (mmol dm^{-3})	7.98	7.29	4.30
Concentration in sap (mmol dm^{-3})	97.8	38.1	30.9
Accumulation ratio	12.3	5.2	7.2

The effect of various factors on accumulation in the vacuole

Temperature

Hoagland and Broyer (1936) observed a marked effect of temperature on accumulation of both cations and anions by their low salt barley roots. Some of their results for potassium are shown in Fig. 3.1. They observed a Q_{10} over the temperature range 6–30°C of 1.6 and over the range 1–24°C it was 1.8. Similar results were obtained by Jacobson, Overstreet and Carlson (1957).

Oxygen partial pressure

The availability of oxygen has been found to be a very important factor in regulating ion uptake by the root. This

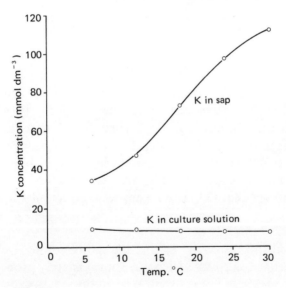

Fig. 3.1

Effect of temperature on potassium accumulation by low salt barley roots over a 10h period. After Hoagland and Broyer (1936).

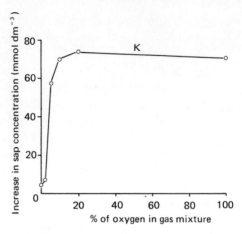

Fig. 3.2

Relationship between oxygen partial pressure in the external solution and accumulation of potassium by excised barley roots. After Hoagland and Broyer (1936).

was clearly shown by Hoagland and Broyer in a number of experiments. They could reduce uptake of K, NO_3 and halides to almost zero if they bubbled nitrogen instead of air through the culture solution. They found that when the percentage of oxygen in the gas mixture was reduced below 10 per cent there was a marked decline in the ability of the roots to accumulate ions (Fig. 3.2). Gas mixtures containing up to 20 per cent CO_2 had no apparent effect on salt uptake.

Carbohydrate levels

Hoagland and Broyer noted that the carbohydrate level in the root declined in a way which was approximately proportional to the increase in salt concentration in the sap. In experiments carried out at about the same time as those of Hoagland and Broyer, Prevot and Steward (1936) found that accumulation of Br by segments of barley roots low in sugar could be

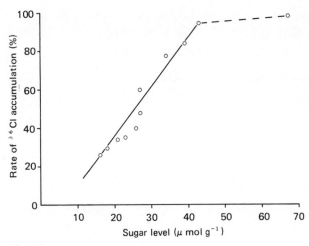

Fig. 3.3

Uptake of ^{36}Cl from 10 mmol dm^{-3} KCl by barley roots in relation to the level of hexose in the cells. Initial ^{36}Cl uptake (100%) = 80 μmol g^{-1} h^{-1}. After Pitman, Mowat and Nair (1971).

accelerated by providing a supply of sugar to the roots in the culture solution. In some more recent experiments Pitman, Mowat and Nair (1971) followed the uptake of ^{36}Cl after low salt barley roots had been transferred to KCl. Accumulation of Cl was at first rapid but fell steadily over a 4 day period. The sugar level in the roots also declined and a close correlation between sugar level and the rate of Cl accumulation was obtained (Fig. 3.3).

The relationship between the rate of salt uptake and endogenous sugar level is however not a simple one and other factors are involved as indicated by the inflection in the curve in Fig. 3.3. Hoagland and Broyer showed that vigorously aerated 'low salt' roots had less sugar but accumulated faster than 'high salt' roots and Cram (1973b) has observed that Cl influx and hexose levels in excised carrot tissue appear to bear no relationship to each other. It is possible that the rate of

35

supply of sugar to the membrane may be a more important factor than the overall sugar level. This might explain some of the conflicting results that have been observed.

The relationship to respiration

The effect of oxygen and carbohydrate levels on the rate of salt uptake implicates respiration in the process. Does respiration have a direct relationship to salt uptake or does oxygen consumption merely maintain cell activity which is only indirectly associated with the accumulation of salt? Hoagland and Broyer took the latter view but Lundegårdh and Burström had earlier obtained evidence which directly implicated respiration in salt uptake. They obtained a close correlation between the uptake of anions from single salt solutions and the evolution of CO_2 from intact wheat roots (Lundegårdh and Burström, 1933; Fig. 3.4). They were able to distinguish two components of respiration; a ground respiration (Rg) and a component which was correlated with anion uptake (KA), sometimes termed salt respiration. Hence, R total $= Rg + KA$. K they termed the anion characterization coefficient as it depended on the species of anion being accumulated. This is illustrated by the differing slopes of the curves in Fig. 3.4. They found that low concentrations of KCN inhibited both anion uptake and KA but Rg was unaffected (Lundegårdh and Burström, 1935). Some of their data are set out in Table 3.2. Their results were confirmed by Robertson and Turner (1945) for discs of carrot tissue and by Handley and Overstreet (1955) using barley roots.

However, although there is no doubt about the phenomenon of salt respiration, the link between salt respiration and the process of salt uptake is still far from clear. Woodford and Gregory (1948) and Handley and Overstreet (1955) found that the ratio of anion uptake/ O_2 absorbed was variable from one experiment to another even when the same salt was used. Another puzzling finding was that the salt

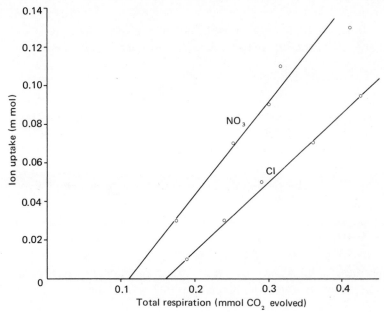

Fig. 3.4

The relationship between NO_3 and Cl uptake and respiration by wheat roots. After Lundegårdh and Burström (1933).

Table 3.2

The effect of KCN on respiration and uptake of potassium and nitrate from 2.5 mmol dm^{-3} KNO_3 by wheat roots. From Lundegårdh and Burström (1935).

KCN (mmol dm^{-3})	K^+ uptake (mmol)	NO_3 uptake (mmol)	Rt (mmol CO_2)	Rg (mmol. CO_2)	KA
0.012	0.176	0.189	0.560	0.182	0.378
0.060	0.145	0.173	0.578	0.232	0.346
0.120	0.083	0.124	0.587	0.339	0.248
0.240	0.115	0.088	0.480	0.304	0.176
0.600	0.059	0.031	0.314	0.252	0.062
3.600	0.013	0.009	0.348	0.330	0.018

respiration was found to persist after removal of the external salt solution. They also found some evidence that cations as well as anions appeared to stimulate respiration.

Effect of the external salt concentration

The rate of salt uptake is dependent upon the external salt concentration, at least in low salt roots. The uptake curve with increasing external concentration can be described by a Langmuir absorption isotherm where the rate of uptake becomes virtually independent of the external concentration (Fig. 3.5). This type of relationship has been found for a wide variety of species (Lyclama, 1963; Tromp, 1962; Hooymans, 1964). However, it is noteworthy that Olsen (1950) failed to find such a relationship in experiments with rye grass. Uptake of NO_3 and PO_4 from vigorously aerated culture solution at constant pH was independent of the external concentration down to 0.003 m mol dm^{-3}.

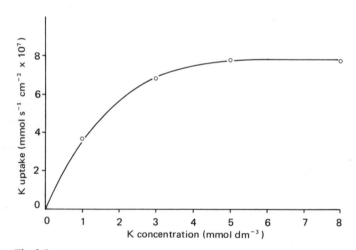

Fig. 3.5

Relationship between potassium absorption by excised barley roots and potassium concentration in the culture solution. After Overstreet, Broyer, Isaacs and Delwiche (1942).

Low salt roots become high salt roots after only 24h in a salt solution. As the ion content of the root rises the net rate of accumulation falls until there is a balance between rate of transport of salt into the cells and salt efflux. The decreased net uptake is partly due to increased efflux but mainly to decreased influx (Pitman, Courtice and Lee, 1968; Johansen, Edwards and Loneragen, 1970). Therefore, with high salt roots at the steady state, decreasing the external concentration should have relatively little effect on the rate of uptake and this may explain Olsen's results. The reason for the decrease in influx is unknown but it is likely to be due to a negative feedback in the uptake mechanism. Pitman, Courtice and Lee (1968) and Pitman, Mowat and Nair (1971) suggested that sugar levels in the cytoplasm may be an important controlling factor. However, Cram (1973b) found that the internal concentration of reducing sugar in carrot and barley root cells showed no correlation with the uptake of Cl, indicating that a decrease in sugar content cannot be the primary cause of the decrease in Cl influx during accumulation. He observed that Cl influx was not correlated with the vacuolar concentration of K, Na and malate, nor with the cellular hydrostatic pressure. There was however a highly significant correlation between Cl influx and log ($Cl + NO_3$ concentration in the vacuole) (Fig. 3.6). Cram suggests three possible explanations for the mechanism of this negative feedback system.

(A) Accumulation of Cl slows down as the Cl electro-chemical potential gradient builds up.

(B) There could be a direct allosteric effect of Cl and NO_3 on the Cl uptake system or on its energy supply.

(C) Internal concentrations of Cl and NO_3 could affect the supply of hexoses to the transport sites in the membrane.

These experiments of Cram are an important contribution and they point the way to further research into the control mechanisms operating in salt uptake.

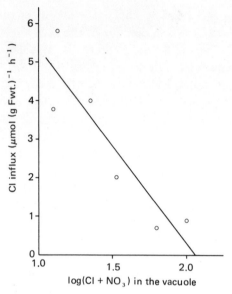

Fig. 3.6

Relationship between Cl^- influx and the log of the internal concentration of $(Cl^- + NO_3^-)$ for barley roots. After Cram (1973b).

pH

The pH of the external solution has been shown to markedly affect the rate of ion uptake. Some early workers found little or no effect of pH (Hoagland and Broyer, 1940; Arnon, Fratzke and Johnson, 1942) but later work showed that anion uptake was reduced at high pH and cation uptake reduced by acid pH values in the external solution. Van den Honert and Hooymans (1955) studied the uptake of nitrate by intact wheat plants. They observed that an increase in pH from 6.0 to 7.4 resulted in a 50 per cent reduction in nitrate uptake as shown in Fig. 3.7. This effect was independent of the external concentration of NO_3. Phosphate uptake was

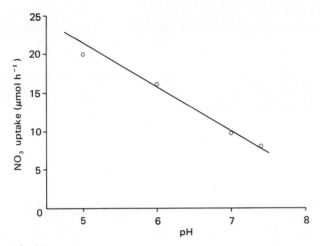

Fig. 3.7

Relationship between pH and nitrate uptake by wheat roots. External concentration of NO_3 0.16 mmol dm^{-3}. After Van den Honert and Hooymans (1955).

affected in a similar way. Olsen (1953) observed that K uptake by rye grass was maximal over the pH range 5 – 8 but there was a decrease below pH 5.0 (Fig. 3.8). In this range inhibition of uptake could be decreased by increasing the K concentration and Olsen thought that K was competing with H$^+$ at low pH levels.

Ulrich (1941) showed that the usual excess of inorganic cations over inorganic anions in the vacuolar sap could be explained by the presence of large quantities of organic acids. Hurd and Sutcliffe (1957) found an increase in uptake of potassium and sodium at pH values above 7.0. Using ^{14}C labelled bicarbonate they showed that this ion was accumulated in equivalent amounts to Na. Bicarbonate, once in the cell, was converted to organic acids such as malic. It appears therefore that the differences between anion and cation uptake are made up by the bicarbonate ion at high pH and by H$^+$ at low pH.

41

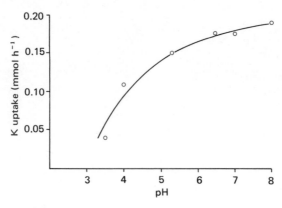

Fig. 3.8

The effects of pH on the uptake of K^+ from
K_2SO_4 (0.4 mmol dm^{-3}) by 4 rye plants. After
Olsen (1953).

Interactions between ions

The rate of uptake of cations from single salt solutions is
dependent upon the nature of the accompanying anion. For
instance, the rate of uptake of potassium by excised barley
roots from KCl is more rapid than from K_2SO_4 (Hooymans,
1968). This effect may be due to the differing permeabilities
of different anion species in the membrane or because some
anions are metabolized and a sink for them is created in the
cell. Blevins, Hiatt and Lowe (1974) found that potassium
uptake from KNO_3 was greater than from KCl in barley
seedlings in the light. In the dark potassium uptake from
KNO_3 was the same as from KCl. They found that nitrate
reductase was induced in the light, and also that the pH of
the cell sap rose when nitrate was accumulated in the light,
which was accompanied by organic acid synthesis. They
suggest that reduction of NO_3 increases the cell pH, and this
in turn increases the synthesis of malic acid. The malic acid
would then provide counterions for increased potassium
uptake.

A number of cations have been found to mutually affect each other. Tromp (1962) found that K^+ absorption by wheat roots was inhibited by NH_4^+ which also had a similar effect on the uptake of Na^+. The uptake of sodium by barley roots in inhibited by potassium (Sutcliffe, 1956; Bange, 1959; Pitman, 1965, 1966) and vice versa (Bange, Tromp and Henkes, 1965).

It is now recognized that calcium plays an important role in the uptake of other ions. F. G. Viets (1944) found that calcium increased the rate of uptake of potassium and bromide by excised barley roots by 100 per cent or more when present up to the ratio 30 Ca/1 K. With higher ratios K absorption was depressed (Fig. 3.9). Mg, Sr, Ba and Al were also effective in stimulating KBr uptake, although to a lesser extent. This stimulating effect by divalent cations has been

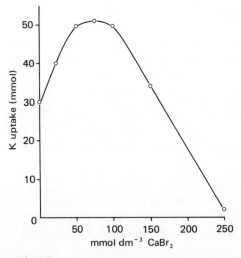

Fig. 3.9

Effect of calcium bromide on the uptake of potassium from KBr (5×10^{-3} mol dm^{-3}) by excised barley roots. After Viets (1944).

43

observed by many workers since then (e.g. Tanada, 1955; Hanson, 1957; Epstein, 1961) and has become generally known as the Viets effect.

The calcium effect does not appear to be related to the effect of calcium on growth as Edwards (1968) found that he could separate the Viets effect from the growth effect in *Trifolium subterraneum.* Kahn and Hanson (1957) suggested that Ca increased the affinity between K and binding sites on the cell membrane whilst Tanada suggested that a ribonuclear protein may be involved. However, Hirata and Mitsui (1965) obtained evidence which showed that the role of calcium is not directly associated with RNA metabolism. Whilst the mechanism of the Viets effect is not understood at the present time it is generally accepted that calcium is essential for the maintenance of the functional integrity of the cell membrane.

Interactions between anions have also been observed. The absorption of phosphate is stimulated by ammonium (Arnon, 1939) whilst ammonium inhibits nitrate uptake (Lyclama, 1963). Interactions between nitrate and bromide and between chloride and bromide have also been demonstrated (Epstein, 1953). The mutual inhibitory interactions have been explained in terms of competition between ions of similar charge for the same binding site on the membrane. The mechanism of the stimulatory effects is however unknown.

The effect of age

The cells of the root become progressively older with increasing distance from the tip as a result of the activity of the apical meristem. Prevot and Steward (1936) were the first to attempt to correlate ion uptake with the growth and development of the root. They found that low salt barley roots exposed to potassium bromide, accumulated bromide at progressively decreasing rates from 0.5 cm behind the tip to the oldest tissues examined at 6.0 cm from the tip

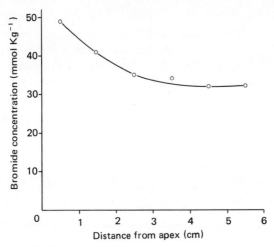

Fig. 3.10

Distribution of bromide in barley roots after
exposure to KBr for 24h. After Prevot and
Steward (1936).

(Fig. 3.10). Similar results were obtained for rubidium by
Steward, Prevot and Harrison (1942) and with ^{32}P for pea
and cotton by Canning and Kramer (1958). Variations on
this relatively simple pattern have also been observed. Kramer
and Wiebe (1952) found considerable variation in the
distribution of ^{32}P even in roots grown in identical conditions. Some roots showed considerable uptake at the tip,
others showed relatively little uptake by the tip, but
considerable uptake further back. The most common condi-
tion they observed was a high accumulation at the tip, a
decrease immediately behind the tip and another peak
6 − 20 mm behind the tip. Rovira and Bowen (1968)
observed two peaks of activity in wheat roots after uptake of
labelled PO_4,Cl and SO_4. The first peak was 4 − 7 mm and
the second 20 − 30 mm from the apex. Eshel and Waisel
(1972) observed a single hump in radioactivity after labelled
sodium uptake by maize roots.

Gregory and Woodford (1939), Machlis (1944) and Berry and Brock (1946) determined the rate of respiration along the root and found it to decline from apex to base. This suggested that the rate of cell metabolism declined steadily with age. However, Brown and Broadbent (1950) in an extremely careful and useful study of the developement of the pea root, showed that on a cell basis, respiration was relatively low at the apex and rose steadily to a maximum at 5 mm behind the apex and thereafter remained fairly constant (Fig. 3.11).

It is worth pointing out that most studies of ion accumulation by different zones of the root have been carried out with salt depleted roots or by using ions not previously present in the root. In roots which have been grown in solutions of uniform concentration for long periods there is very little variation in the steady state vacuolar concentration from apex to base as the results of Pitman and

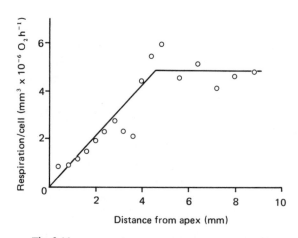

Fig. 3.11

Average cell respiration at increasing distance from the apex of pea roots. After Brown and Broadbent (1950).

Table 3.3

Potassium and sodium concentrations along barley roots in flux equilibrium. From Pitman and Saddler (1967).

Distance from apex (mm)	Na in root (mmol (g Fwt.)$^{-1}$)	K in root (mmol (g Fwt.)$^{-1}$)
0–2	11	85
2–6	15	78
6–10	17	88
10–14	18	92
14–24	19	92
24–34	18	86

Saddler (1967) show for barley (Table 3.3). Nevertheless the peaks observed by a number of workers at about 5 mm and at 20–60 mm from the apex need to be explained and it seems likely that with increasing age changes occur in the permeability of the cell membrane and perhaps also in the mechanism for the transport of the ions into the cell. In this connection Eshel and Waisel (1973) have shown that the degree of inhibition caused by rubidium on sodium uptake varies along the root. They found that inhibition of sodium uptake near the apex was relatively small compared to that at a distance of 6 − 8 cm from the tip.

The effect of micro-organisms

Plants growing in soil are surrounded by large numbers of micro-organisms. On average most soils contain about 2×10^9 bacteria/gram. The root surface itself is colonized by bacteria and fungi. In many cases there is a close association between the root system and a particular species of fungus. The plant and the fungus form a symbiotic relationship termed a mycorrhiza. The discoverer of mycorrhiza in the last century, Frank, was also the first to recognize their importance in plant nutrition, He observed that the presence of the fungus enhanced plant growth. The nutritional importance of soil and root colonizing bacteria has only recently been considered.

Mycorrhiza

Mycorrhiza may be divided into two kinds; ectotrophic and endotrophic. The ectotrophic mycorrhiza have a fungal sheath which completely encloses the root. They are commonly found in many forest trees such as pine, spruce, oak and beech (Plates 1 and 3). A well developed sheath is about 40 μm thick. Endotrophic mycorrhiza do not form an external mantle of fungus but the fungus lives inside the root, both between and within the cells. There are hyphal connections between the mycelium in the root and a loose reticulum of hyphae in the surrounding soil. Endotrophic mycorrhiza are found in most families of angiosperms as well as in some gymnosperms and cryptogams.

The presence of the fungal sheath of the ectotrophic mycorrhiza has been found to enhance salt uptake. Hatch (1937) found in experiments with roots of *Pinus strobus* that

Table 3.4

The effect of mycorrhiza on growth and nutrient uptake by *Pinus strobus*. From Hatch, 1937.

	Mycorrhizal	Non-mycorrhizal
Dry weight of seedlings (mg)	404.6	320.7
Nitrogen content percent of dry weight	1.24	0.85
Phosphorus content % DWt.	0.196	.074
Potassium content % DWt.	0.74	0.43

mycorrhizal roots took up greater amounts of potassium, nitrogen and phosphorus than roots without the fungal symbiont. It can be seen from Hatch's data in Table 3.4 that phosphorus uptake was conspiciously increased. The uptake of phosphate has been particularly well studied as it is only in phosphate deficient soils that mycorrhizal infection is effective in increasing the growth of the host.

Harley and McCready (1952) found that $80 - 90$ per cent of the phosphate taken up by beech mycorrhiza remained in the fungus. This could later be transferred to the root under certain circumstances such as if the mycorrhiza was deprived of an external supply of phosphate (Harley and Brierley, 1954). Later work by Harley and his colleagues (Harley, 1969) has shown that four main processes appear to control phosphate uptake by ectotrophic mycorrhiza:

(a) Uptake by the fungus.

(b) Transfer from the fungus to the host.

(c) Diffusion through the fungal sheath via cell walls and intercellular spaces.

(d) Uptake by the root of phosphate arriving by pathway (c).

At low concentrations (below 0.3 mmol dm^{-3}) the sheath appears to take up all the available phosphate and so prevent diffusion directly to the root. With increasing external concentrations phosphate is able to reach the root by processes (c) and (d).

The commonest type of endotrophic mycorrhiza is that formed by plants with a number of fungi of the phycomycetes. Most of these fungi belong to the genus *Endogone*. It has only recently become apparent that mycorrhiza formed with *Endogone* are widespread and may be found in many commercial crops such as rubber, tea, tobacco and cereals. At the present time however we know little about the biology of these mycorrhiza and virtually nothing about their nutrition.

Like the ectotrophic mycorrhiza their presence does increase plant growth. It has been found by Gerdemann (1961) and Daft and Nicholson (1966) that plants infected with *Endogone* show better growth and greater ability to absorb phosphate than uninfected plants. The influence of the fungus was most pronounced in conditions of low external phosphate. Furthermore Nicholson (1967) has shown that tomato plants infected with *Endogone* can derive phosphate from compounds of phosphorus which are of no nutritive value to uninfected plants. It appears that the hyphae of *Endogone* can exploit the soil for phosphate much more efficiently than the root.

Bacteria and non-mycorrhizal fungi

The study of the effect of the general soil microflora on salt uptake by roots has centred on the uptake of phosphate. This is because phosphate occurs in the soil solution at very low concentrations and micro-organisms have a great avidity for it. In order to separate the effect of the root from the effect of the attendant microflora, experiments have been carried out with plants grown in sterile or non-sterile conditions in culture solution or nutrient agar.

Uptake of ^{32}P by tomato and clover roots was investigated by Bowen and Rovira (1966). They found that uptake by non-sterile roots was up to 85 per cent greater than by the corresponding sterile roots. Transport to the shoot was also greater in the non-sterile plants. Bacteria on the surface on the non-sterile roots were presumably responsible for the higher labelled phosphate uptake but their presence cannot explain the increased transport to the shoot. Barber, Sanderson and Russell (1968) provided visual evidence of the effects of bacterial contamination with autoradiographs of labelled phosphate distribution in barley roots. While sterile roots

Table 3.5

Uptake and distribution of labelled phosphate during a 24h period in barley grown under sterile and non-sterile conditions. From Barber and Loughman (1967).

Treatment	Total P uptake µg/g Dwt.	% transferred to shoot
0.001 ppm P sterile	4.8	20.4
0.001 ppm P non-sterile	2.6	2.3
10 ppm P sterile	3062	74.6
10 ppm P non-sterile	2525	78.8

showed uniform distribution of ^{32}P within the root, the presence of micro-organisms caused considerable concentration of phosphorus at the root surface. Barber and Loughman (1967) found that with low external phosphate concentration (0.001 ppm P) there was much more phosphate transported to the shoots in sterile plants than in non-sterile plants (Table 3.5). This is what might be expected if the bacteria on the root surface compete with the root for phosphate. Barber (1968) has suggested that the results of Rovira and Bowen (1966) may have been due to their roots having a wide spectrum of species producing stimulating secretions which may predominate over the competition effect.

Micro-organisms therefore appear to have a great influence on phosphate uptake by roots from low external concentrations. There is very little effect at high phosphate concentrations (Barber and Loughman, 1967). The influences of micro-organisms on the uptake of other ions has not been studied.

4 Active transport

Criteria for active transport

The realization by Hoagland and his contemporaries that uptake of solutes by plant cells is not controlled merely by physical processes but is closely connected with metabolic activity gave rise to a new concept and a new term — active transport. This concept was subsequently extended to salt uptake in animal cells and it has dominated the study of solute accumulation by cells and tissues for the past 40 years. A large amount of data has been obtained over this time which indicates that salt uptake is a distinct biological process which at present cannot be explained in terms of physics and chemistry or biochemistry. Much of our information however is circumstantial and we know relatively little about what goes on at the hub of the active transport process, consequently it has proved rather difficult to define. The use of the term active transport has therefore become somewhat controversial in some quarters.

Early investigators tried to relate transport of solutes to specific metabolic processes. This is a legitimate aim, but with plant cells it has met with only limited success so far.

From time to time various workers have put forward a number of critieria as evidence for active transport such as:

(a) Accumulation against a concentration gradient.

(b) Inhibition of absorbtion by metabolic inhibitors such as cyanide.

(c) Competition for entry into the cell between different solutes.

(d) The rate of absorption not being a linear function of the concentration difference between the outside and inside of the cell.

(e) A high temperature coefficient (Q_{10}).

Each one of these criteria has weaknesses. Criterion (a) is satisfactory for the uptake of non-electrolytes like sugars. For electrolytes where there is an electrical potential gradient it is possible to have accumulation of an ion against the concentration gradient without active transport. For instance if there are large non-diffusing molecules with negative charges in the cytoplasm a Donnan system will develop resulting in a higher concentration of diffusable cations in the cytoplasm than outside. If the product of the concentrations of cations and anions inside is greater than that outside then of course this type of accumulation cannot be explained in terms of a Donnan equilibrium. The important point is that accumulation of ions against a concentration gradient cannot be used as a definition of active transport.

Inhibition of ion transport by inhibitors of metabolism is a valid way for testing for active transport provided that the inhibitor is directly affecting the transport process. Unfortunately the specific action of most metabolic inhibitors is not precisely known in a complex tissue like the root. There are two main disadvantages with inhibitors, firstly they frequently appear to affect salt uptake indirectly by inhibiting a process remote from the active transport itself and secondly they are not usually specific but may affect several processes simultaneously. For example the antibiotic cyclo-heximide is an inhibitor of protein synthesis in fungi. In roots it also inhibits protein synthesis (Läuchli, Luttge and Pitman, 1973), but Macdonald and Ellis (1969) found that it stimulates respiration in a number of root tissues indicating that it can also interfere with energy transfer.

Competion and non linearity of uptake rate with increasing external concentration both suggest that the ion becomes attached to some substance in the membrane. These phenomena will be dealt with in more detail in the next chapter as they are an important facet of active transport. However, they are not always accompanied by active transport. In the red blood cell, various monosaccharides compete for sites in the membrane which show saturation at high external sugar concentrations. This transport, however, appears to be a purely physical process as the sugar concentration inside the cell never exceeds that outside (Willbrandt, 1954).

Physical processes like diffusion are often rather insensitive to temperature, a rise in temperature of $10°C$ usually increases diffusion by something less than 20 per cent (i.e. Q_{10} of approximately 1.2). Metabolic reactions however show greater dependence on temperature and usually show a high Q_{10} and observation of a high Q_{10} for salt uptake (Fig. 3.1) has sometimes been interpreted as indicating active transport. Physical theory of kinetic energy indicates, however, that any process having an appreciable energy barrier can have a high temperature coefficient. A membrane may provide a formidable energy barrier and diffusion across such a membrane will have a high Q_{10} (Nobel, 1970). Thus a high Q_{10} does not necessarily imply active transport. Some experiments of Jackson and Weatherley (1962) clearly illustrate this point. They artificially induced a water flow across excised tomato roots using hydrostatic pressure. Reducing the ambient temperature caused a marked reduction in the flux of water. Over the temperature range

$9-19.5°C$ their data give a Q_{10} of 3 yet there was no evidence that active transport of water was involved in their experiments. A lucid account of the physical chemistry of temperature coefficients is given by Nobel (1970).

Taken individually each of the above criteria is obviously inadequate as a basis for active transport but an uptake process with several of these characteristics is likely to be active as Jennings (1963) has pointed out. Nevertheless a much more reliable criterion for active transport is required. Generally speaking, cells and tissues engaged in active transport are not in equilibrium conditions. Water, ions and other solutes moving through a membrane can exert a frictional drag on each other, thus the fluxes of various solutes across membranes are interdependent. Irreversible thermodynamics provides a quantitative description of these fluxes.

It has been found that the flow of substances may depend in a direct and linear manner not only on the conjugated force (e.g. electrical potential difference is the force to which current flow is conjugated) but also on non-conjugate forces (e.g. temperature may induce a current flow in some systems). To cover all possibilities, the flows or fluxes, J_1 and J_2 of two substances, 1 and 2, whose conjugate forces are X_1 and X_2 would have to be written

$$J_1 = L_{11}X_1 + L_{12}X_2 \qquad (4.1)$$

$$J_2 = L_{21}X_1 + L_{22}X_2. \qquad (4.2)$$

The coefficients L_{11}, L_{12}, L_{21}, and L_{22} are called phenomenological coefficients and in general are non-zero. For additional substances the equations may be expanded. For n substances there will be n^2 phenomenological co-efficients and,

$$J_i = \sum_{k=1}^{n} L_{ik}X_k \qquad (4.3)$$
$$(i = 1, 2, 3 \dots n).$$

55

The large number of independent coefficients is reduced by Onsager's law, applicable under some conditions which states:

$$L_{ik} = L_{ki}. \tag{4.4}$$

Because of the linear nature of equations (4.1) and (4.2) it is possible to transform them into

$$X_1 = R_{11}J_1 + R_{12}J_2 \tag{4.5}$$

$$X_2 = R_{21}J_1 + R_{22}J_2 \tag{4.6}$$

or in the general case,

$$X_i = \sum_{k=1}^{n} R_{ik}J_k. \tag{4.7}$$

The coefficients R_{ik} have the dimensions of force per unit flow and therefore are generalized resistances or frictions.

Kedem (1961) has attempted to define active transport in terms of irreversible thermodynamics and she has suggested the following equation to account for fluxes in biological systems

$$\Delta\bar{\mu}_i = \sum_{k=1}^{n} R_{ik}J_k + R_{ir}J_r. \tag{4.8}$$

In this equation J_r is the rate of some biochemical reaction and R_{ir} represents the extent of coupling of that reaction to the movement of species i against an electrochemical potential gradient.

Thus active transport is envisaged as an entrainment between a transport flux and a metabolic reaction. In the language of irreversible thermodynamics active transport is characterized by a non-zero coupling coefficient between the flow of substance and the metabolic reaction.

Unfortunately just what biochemical process J_r represents is a matter of speculation. It could be a particular enzymic reaction and this will be considered in more detail in Chapter 6. MacRobbie (1970) has pointed out that a definition of active transport based on irreversible thermodynamics, although a rigorous one, is unsatisfactory because most of the coupling or phenomenological coefficients which are the currency of irreversible thermodynamics are at present impossible to measure. In fact, for a single cell in potassium chloride solution fifteen coefficients would be needed! Thus for the time being the application of irreversible thermodynamics has outstripped the technical ability of the plant physiologist.

A very useful definition of active transport was put forward in 1949 by the Danish physiologist H. H. Ussing. He defined active transport as a process by which an ion is moved against an electrochemical potential gradient. An ion will move spontaneously down a gradient of electrochemical potential and movement against such a gradient is dependent on a decrease in free energy of some metabolic process. Ussing's definition, although not as rigorous as the definition above based on irreversible thermodynamics, is a very important one because it can be tested by experiment.

In plant roots the most successful way of detecting active transport of ions has been by the use of the Nernst equation. This describes the simple relationship between the electrical potential difference and the distribution of ions across a membrane at equilibrium.

Nernst potential

The chemical potential or relative free energy of an ion j is given by:

$$\bar{\mu}_j = \bar{\mu}_j^* + RT \ln a_j + \bar{V}_j P + z_j FE. \tag{4.9}$$

Chemical potential is a relative quantity so the term $\bar{\mu}_j^*$ is included in equation (4.9) as a reference point. The term $\bar{V}_j P$

57

represents the effect of pressure on chemical potential. With roots, pressure effects are usually negligible and so this term can be omitted here. Equation (4.9) thus becomes

$$\bar{\mu}_j = \bar{\mu}_j^* + RT \ln a_j + z_j FE. \tag{4.10}$$

where R = the gas constant, T = absolute temperature, a_j = the chemical activity of ion j, z = algebraic valency of ion j, F = the Faraday and E = the electrical potential in volts.

Let us consider a membrane with ion j distributed across it at equilibrium

OUTSIDE | INSIDE

$$\bar{\mu}_j^o = \bar{\mu}_j^* + RT \ln a_j^o + z_j FE^o \quad | \quad \bar{\mu}_j^i = \bar{\mu}_j^* + RT \ln a_j^i + z_j FE^i$$

At equilibrium the chemical potential on the outside equals that on the inside i.e. $\bar{\mu}_j^o = \bar{\mu}_j^i$. Therefore

$$\bar{\mu}_j^* + RT \ln a_j^o + z_j FE^o = \bar{\mu}_j^* + RT \ln a_j^i + z_j FE^i \tag{4.11}$$

and

$$E^i - E^0 = \frac{RT}{z_j F} \ln \frac{a_j^o}{a_j^i} \tag{4.12}$$

Equation (4.12) is known as the Nernst equation and $E^i - E^o$ is the Nernst potential. For practical purposes the right-hand side of equation (4.12) is simplified by converting to common logarithms and including numerical values for RT/F. At $20°C$ equation (4.12) becomes

$$E_{N_j} = \frac{58}{z_j} \log \frac{a_j^o}{a_j^i} \text{ mV.} \tag{4.13}$$

With dilute solutions where the activity coefficient of the ion is close to 1, concentrations are often used in place of the activities. Equation (4.13) applies to cations. For anions the quotient a_j^o/a_j^i is inverted because of the sign change.

Membrane potentials in roots and the application of the Nernst equation

The Nernst equation can be used as a test for active transport as defined by Ussing provided that the concentration (or activity) of a given ion both inside and external to the cell is known. In addition, the electrical potential difference between the inside and outside is required. The Nernst equation can then be used to predict the internal concentration of the ion. For example, if the electrical potential difference is 58 mV with the inside of the cell negative and the external concentration is 1 mmol dm^{-3} then according to the Nernst equation the internal concentration of a given monovalent cation should be 10 mmol dm^{-3}. If the internal concentration is determined and found to be significantly greater than 10 mmol dm^{-3} then this is taken as evidence for active uptake of the ion. It must be remembered however that this sort of analysis is only valid if the cell or tissue is in equilibrium with its surroundings. In non-equilibrium conditions the Nernst equation does not hold and a more complex relationship is applicable. This will be discussed later.

One of the first applications of the Nernst equation to plant cells was made by MacRobbie and Dainty in 1958 on the large celled brackish water algae *Nitellopsis obtusa*. As the cells of this species are about 6 cm long and 700 μm in diameter it is comparatively easy to obtain uncontaminated samples of the vacuolar sap for analysis. They obtained the potential difference between the sap and the external solution by inserting a microelectrode into the vacuole.

The technique for measuring cell PD is now a standard procedure. Glass micropipettes are pulled from glass tubing using an electrode puller. They can be made with tips with an

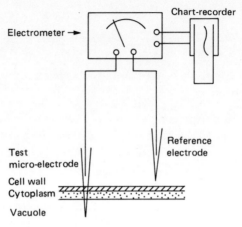

Fig. 4.1

Circuit used for the determination of the
electrical potential difference between
the outside solution and the vacuolar sap.

opening of less than 1 μm in diameter. The micropipettes are
then filled with 3 mol dm^{-3} KCl and connected to either
Ag/AgCl or calomel electrodes which in turn are connected
to an electrometer with a high input impedance. A diagrammatic representation of this circuit is shown in Fig. 4.1. The
test electrode is implanted in the cell with a micromanipulator and the reference electrode is positioned nearby in the
bathing solution as shown in Plate 4. A recorder is usually
connected to the electrometer in order to follow trends in
electrical potential.

MacRobbie and Dainty recorded a steady resting potential
of approximately 150 mV between the vacuolar sap and the
external medium with the sap negative. They found that the
ions K$^+$, Cl$^-$ and Na$^+$ were approximately in flux equilibrium
under experimental conditions. However, calculation of
the Nernst potential showed that there were large electrochemical potential differences between the sap and the

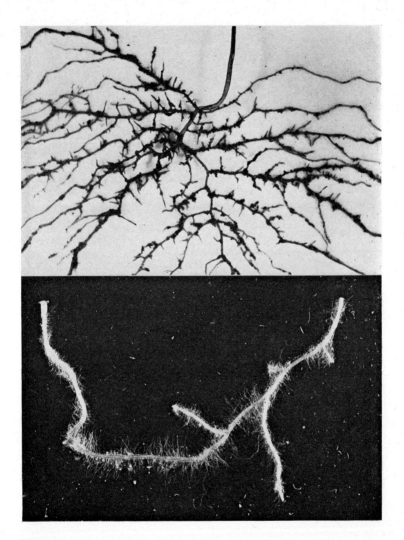

Plate 1
The root system of Sitka spruce (*Picea sitchensis*).
Top: Mycorrhizal tap root showing secondary and tertiary branching.
Bottom: A non-mycorrhizal root with abundant root hairs.
Photographs by Ian J. Alexander.

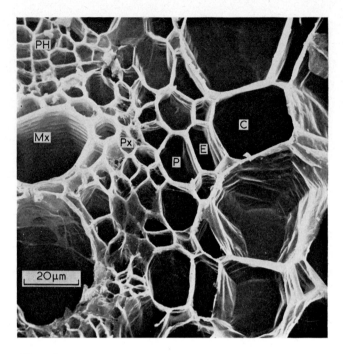

Plate 2
Scanning electron-micrograph of part of a transverse section of the root of sunflower (*Helianthus annuus*) showing the inner cortex and stele. The section was taken approximately 1 cm behind the apex. C, cortical cell, E, endodermis. P, pericycle. Mx, metaxylem vessel. Px, protoxylem vessel. PH, phloem.
Prepared and photographed by Richard P. C. Johnson and David John.

Plate 3 (Top right)
Transverse section of a mycorrhizal root of Sitka spruce showing the fungal sheath.
Photograph by Ian J. Alexander.

Plate 4 (Bottom right)
Micro-electrode implanted in an epidermal cell of the root of *Zea mays* to measure the trans-membrane electrical potential. The reference electrode in the bathing solution can also be seen.
Photograph by James Dunlop.

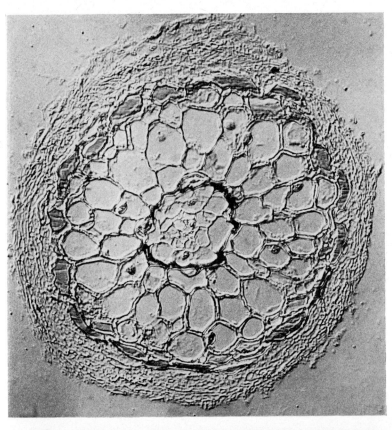

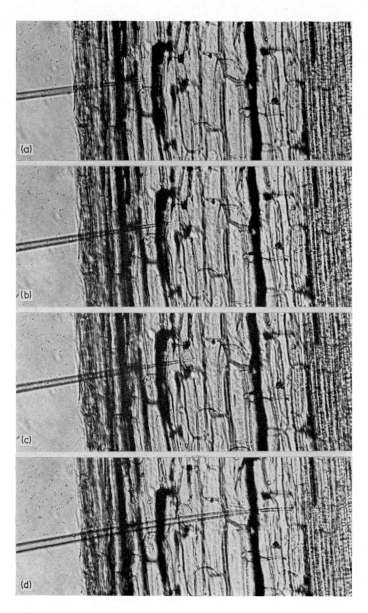

Plate 5
Measurement of the profile of vacuolar potassium activity in the root of *Zea mays* using a potassium specific micro-electrode. In (a), (b) and (c) the electrode is in the 4th, 5th and 6th cortical cells respectively. In (d) it is in the endodermis.
Photograph by James Dunlop.

Table 4.1

Concentrations of Na^+, K^+ and Cl^- in the sap of *Nitellopsis obtusa* and in the external solution together with the ion equilibrium potentials (Nernst potentials). Data from MacRobbie and Dainty (1958).

Ion	Sap concentration (mmol dm⁻³)	External concentration (mmol dm⁻³)	$\dfrac{RT}{zF} ln \dfrac{C_0}{C_i}$ (mV)
Na^+	54	30	-15
K^+	113	0.65	-130
Cl^-	206	35	$+45$

outside solution for Na^+ and Cl^- (Table 4.1). For Cl^- $\bar{\mu}_i > \bar{\mu}_0$, for Na^+ $\bar{\mu}_i < \bar{\mu}_0$ and for K^+ $\bar{\mu}_i \simeq \bar{\mu}_0$. These figures suggest that Cl^- is actively accumulated and that Na^+ is actively extruded by the cells. In the jargon of the physiologist there appears to be a Cl^- influx pump and a Na^+ efflux pump operating. K^+ appears to be in passive equilibrium.

A large number of electropotential measurements have been made on various species of large celled algae but the application of the technique to roots has been hindered by the difficulty of implanting microelectrodes in root cells because of their relatively small size. However, Etherton and Higinbotham (1960) reported measurements of electrical potential difference for the root hair cells of *Avena*. The vacuole was found to be 80 mV more negative than the bathing solution (1 mmol dm⁻³ KCl). The PD was markedly affected by the external concentration of potassium. Increasing the external concentration of potassium tenfold caused a reduction of approximately 50 mV in the PD. This effect appears to be a common characteristic of the membrane of root cells. Dunlop and Bowling (1971a) found a value of 32 mV per tenfold change in the external concentration of potassium in maize roots and Sinyukhin and Vyskrebentseva

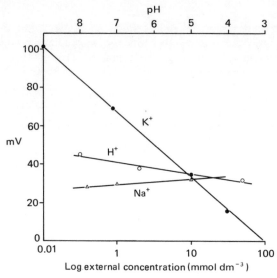

Fig. 4.2

The effect of the external concentration of K^+, H^+ and Na^+ on the cell electrical potential difference of epidermal cells of sunflower roots bathed in nutrient solution. Each point is the mean of ten or more readings. Data of the author.

(1967) obtained a value of 35 mV for pumpkin roots. Other ions such as Na^+ and H^+ however, appear to have much less effect on the cell potential (Fig. 4.2).

Etherton and Higinbotham attempted to ascertain the location within the cell of the potential difference. In the root hair cells of *Avena* at a stage in the root growth a thick layer of cytoplasm is clearly visible and so they were able to see whether the tip of the microelectrode was in the cytoplasm or the vacuole. The potential difference was measured when the electrode was inserted slowly into the cytoplasm and when it entered the vacuole. They observed a large rise in potential when the electrode entered the cytoplasm but little further change when it entered the vacuole. They obtained the following cytoplasm-vacuole readings in 1 m mol dm^{-3} KCl; 83/90. 87/84, 75/75 and 71/71. Thus their data indicate that the cell potential resides almost entirely at the plasmalemma, a situation which is similar to that in certain algae. This is an important conclusion which, because of the technical difficulties of measuring PDs in the thin layer of cytoplasm in root cells, has been difficult to check. A number of workers have subsequently assumed that the PD across the tonoplast is zero for the purposes of calculation but it is obviously important that this assumption should be confirmed by further experiments along the lines pioneered by Etherton and Higinbotham.

The first results of the application of the Nernst equation to roots were obtained by Etherton (1963). He observed potentials of up to 117 mV in the cells of oat and pea roots. Analysis of the tissues for sodium and potassium indicated that sodium appeared to be actively extruded from the cells. Potassium appeared to be either in passive diffusion equilibrium or actively extruded depending upon the external potassium concentration.

The first comprehensive studies of the electro-chemical status of ions in roots were carried out by Bowling, Macklon and Spanswick(1966), Bowling (1966) and Higinbotham, Etherton and Foster (1967).The most important finding from all these experiments was that all the anions investigated were found to be transported and accumulated against the electrochemical potential gradient. The behaviour of the cations on the other hand was not so clear. They appeared to be either actively extruded or in passive equilibrium and in some experiments potassium appeared to be actively transported into the cells. These important conclusions are illustrated in Table 4.2 in which are set out some of the data of Higinbotham *et al.* for pea roots. They found that the concentrations in the tissues of most of the ions changed very

63

Table 4.2

Electrochemical equilibria of the major ions in the root of *Pisum sativum*. Data from Higinbotham, Etherton and Foster (1967).

Ion	External concentration ($mmol\,dm^{-3}$)	Internal concentration ($mmol\,dm^{-3}$)	Internal conc. predicted by Nernst equation ($mmol\,dm^{-3}$)	Conclusion
K^+	1.0	75	74	Passive equilibrium
Na^+	1.0	8	74	Active extrusion
Mg^{++}	0.25	3	2 700	Passive uptake
Ca^{++}	1.0	2	10 800	Passive uptake
NO_3^-	2.0	28	0.0272	Active uptake
Cl^-	1.0	7	0.0136	Active uptake
$H_2PO_4^-$	1.0	21	0.0136	Active uptake
SO_4^{--}	0.25	19	0.00009	Active uptake

little with the time and so could assume that the ions were close to flux equilibrium. The electrical potential measurements (mean value -110 mV) were however confined to the epidermal cells, whilst the internal ion concentrations were obtained from water extracts of the bulk of the root tissue. Despite this discrepancy there can be no doubt from the data that there is active anion transport. There is no evidence for active accumulation of Ca^{++} and Mg^{++} and all the information suggests that these ions move into roots down the electrochemical potential gradient. In the roots of some aquatic species these ions appear to be in passive equilibrium which suggests that they move into the cell solely in response to the electrical gradient (Shepherd and Bowling, 1973).

The behaviour of sodium is rather complex. A number of workers have found that application of the Nernst equation to the roots of various species indicate that sodium is actively extruded (Etherton, 1963, Higinbotham *et al.*, 1967 and

Pitman and Saddler, 1967). However the presence of an outwardly directed sodium pump is by no means universal. Nemcek, Sigler and Kleinzeller (1966) obtained good evidence that the cells in the wall of the pitcher of *Nepenthes henryana* actively accumulate sodium. Also Bowling and Ansari (1971) and Shepherd and Bowling (1973) have obtained evidence which indicates that the roots of a number of species are able to actively accumulate sodium when the external concentration is low. This behaviour may be related to the fact that some species show growth responses in the presence of low concentrations of sodium (Montasir, Sharoubeem and Sidrak, 1966). As sodium generally inhibits growth when present at high concentrations and is beneficial at low concentrations there is probably some mechanism for keeping the internal concentration to within fairly precise limits and this may account for the divergent results that have been reported.

Potassium has been found to be actively extruded (Etherton, 1963) or in passive equilibrium (Higinbotham *et al.*, 1967; Pallaghy and Scott, 1969), or actively accumulated (Pitman and Saddler, 1967; Bowling and Ansari, 1971). In reviewing this situation Higinbotham (1973) is inclined to the view that a clear case for active potassium influx in higher plants has not yet been made, in contrast to the evidence for algae. It is certainly true that the differences between the measured internal concentrations of potassium and those predicted by the Nernst equation are usually not very great. On the other hand it can be argued that as potassium is so mobile in plant tissues passive diffusion could be masking the activity of a potassium pump. In this case the use of the Nernst equation would be an inappropriate test for active transport.

In animal cells there is well documented evidence for a pump in which active efflux of Na^+ is linked with active K^+ influx and which is dependent upon ATPase activity

(Whittam and Wheeler, 1970). Na^+ efflux is dependent upon the external potassium concentration and is inhibited by the cardiac glycoside ouabain. A K/Na pump of this type also appears to occur in some algae (MacRobbie, 1970). Attempts have been made to detect a similar system in the cells of higher plants. Jeschke (1970) has reported evidence for a potassium stimulated sodium efflux in barley roots and Krasavina and Vyskrebentseva (1972) and Nassery and Baker (1972) have observed that ouabain inhibits sodium efflux from *Cucurbita* and barley roots. However Nassery and Baker found that potassium inhibited sodium efflux and Jeschke (1973) and Pitman and Saddler (1967) observed little effect of ouabain on sodium efflux in barley. Such conflicting evidence and also the uncertainty about whether there is active potassium transport has led Higinbotham (1973) to suggest that a K/Na linked transport system does not occur in the cells of higher plants.

The electrochemical state of important ions like $^-HCO_3$ and H^+ has been little studied in roots. Raven and Smith (1973) have pointed out that according to the Nernst equation the potential difference across the plasmalemma of plant cells (usually about -100 mV) implies a cytoplasmic pH which is more acid than that of the outside solution if H^+ is in flux equilibrium. Therefore to keep the cytoplasmic pH at approximately 6.0–7.0 which is the optimum for most enzymes, extrusion of H^+ against the electrochemical potential gradient must occur. Thus circumstantial evidence suggests the presence of a H^+ efflux pump in root cells. There is however very little direct evidence for this at present but what evidence there is does suggest active H^+ efflux. Pitman (1970) observed an efflux of H^+ from cells of low salt barley roots during sodium accumulation. This H^+ loss appeared to be active as it occurred against the electrochemical potential gradient and was inhibited by inhibitors of oxidative phosphorylation. Although it is difficult to measure the pH of the

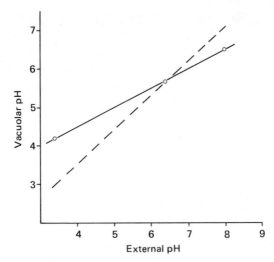

Fig. 4.3

The relationship between the external pH and the pH of the vacuolar sap of cells of the root of *Helianthus annuus*. The broken line shows the predicted vacuolar pH assuming H^+ to be in passive equilibrium. (From Bowling, 1974).

cytoplasm of root cells the vacuolar pH can be determined with reasonable precision with H^+ specific microelectrodes. If the cell membrane potential is also measured then the behaviour of H^+ can be determined from the application of the Nernst equation. With low external pH the vacuolar pH of sunflower root cells was found to be much higher than predicted (Fig. 4.3) suggesting that H^+ was being pumped out of the cell against the electrochemical potential gradient (Bowling, 1974).

Origin of the membrane electrical potential difference

The membrane potential is greatly influenced by the concentration of ions in the external solution as shown by the curves in Fig. 4.2. Potassium has the greatest effect, depolarizing the potential as the external concentration is increased. Some ions however, have the reverse effect. Higinbotham, Etherton and Foster (1964) found the PD of the cells of *Avena* coleoptiles to be hyperpolarized by Ca^{++} and H^+. The depolarizing effect of potassium can be explained in terms of a diffusion of the ion from the relatively high concentration in the cytoplasm across the membrane, which is permeable to K^+, into the relatively low external concentration. Increasing the external concentration will reduce diffusion of potassium from the cell and hence reduce the diffusion potential.

Diffusion potentials depend on the difference in permeability of the membrane to different ions. Let us consider a semi-permeable membrane which has a higher concentration of a salt, say KCl, on the inside than on the outside.

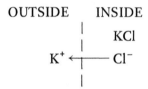

If the membrane is permeable to K^+ but not to Cl^-, then K^+ will diffuse across the membrane from right to left until equilibrium is attained. At equilibrium the inside of the membrane will be negative in relation to the outside because of the preponderance of Cl^- over K^+ on the inside. If the concentration of KCl on the inside is 10 times that on the outside then at $20°C$ according to the Nernst equation the potential difference across the membrane will be 58 mV. In practice Cl^- is not completely impermeable and so the membrane potential never attains the theoretical maximum value in roots. Also a number of other ions will be permeating the membrane and the potential will be affected both by their charges and their permeabilities.

The diffusion potential across a membrane can be described by an expression derived by Goldman (1943).

$$E_m = \frac{RT}{F} \ln \frac{\Sigma P_{j^+} C_j^o + \Sigma P_{j^-} C_j^i}{\Sigma P_{j^+} C_{j^+}^i + \Sigma P_{j^-} C_j^o} \qquad (4.14)$$

where P is the permeability coefficient of a univalent cation j or univalent anion j and C is the activity of the anion or cation. The other notations are as before.

For algae growing in sea water where only three major ions are involved namely K^+, Na^+ and Cl^- the following expression for the Goldman equation has been employed.

$$E_m = \frac{RT}{F} \ln \frac{P_K C_K^o + P_{Na} C_{Na}^o + P_{Cl} C_{Cl}^i}{P_K C_K^i + P_{Na} C_{Na}^i + P_{Cl} C_{Cl}^o}. \qquad (4.15)$$

In some species of algae with large cells it is possible to remove the cytoplasm and analyse it for ions. For example Spanswick and Williams (1964) were able to extract the cytoplasm from the cells of the fresh water species, *Nitella translucens* by centrifuging the cytoplasm to one end of the cell when it could be removed quite easily for analysis. Their results are summarized in Table 4.3. The potential difference across the plasmalemma of cells growing in artificial pond

Table 4.3

Cytoplasmic concentrations of *Nitella transluscens* grown in artificial pond water. Data from Spanswick and Williams (1964)

Ion	Artificial Pond Water ($mmol\,dm^{-3}$)	Flowing cytoplasm ($mmol\,dm^{-3}$)
Na^+	1.0	14
K^+	0.1	119
Cl^-	1.3	65

water was approximately -138 mV. In some earlier work on *Nitella translucens*, MacRobbie (1962) obtained data on the relative permeabilities of K^+ and Na^+ through the plasmalemma. Taking P_K as 1, P_{Na} was found to be 0.18. Nobel (1970) has used the data of Spanswick and Williams and MacRobbie to solve the Goldman equation for the membrane potential assuming P_{cl} to be 0.003. The value obtained is -140 mV which is in good agreement with the measured value. This strongly supports the view that the membrane potential is a diffusion potential. The same conclusion was reached by Black and Weeks (1972) who performed a similar exercise on the alga *Enteromorpha*. Unfortunately we do not have this sort of information for roots because of the difficulty of determining cytoplasmic concentrations.

In photosynthetic tissues however, there is increasing evidence for a component of the membrane potential which is not explicable in terms of diffusion. It seems highly likely that active transport systems may make a contribution to the membrane potential by the transfer of charges across the membrane. Such transport is termed 'electrogenic' and is described as being brought about by 'electrogenic pumps'. The best criterion for the presence of an electrogenic pump is the generation of a membrane potential which is greater than that predicted by the Goldman equation.

Spanswick (1972) has obtained evidence for a light stimulated electrogenic pump in *Nitella translucens*. In the dark he observed that the membrane potential became equal to the calculated K^+ equilibrium potential but in the light (1.0 mW cm^{-2}) the membrane potential was hyperpolarized by 50 mV. He subsequently concluded that this electrogenic pump is due to active transport of H^+ powered by ATP (Spanswick, 1974) and also observed a similar light induced hyperpolarization of the membrane potential in the water plant *Elodea canadensis* (Spanswick, 1973).

Evidence for electrogenic pumps in root cells is meagre.

This is because, as already mentioned, it is difficult to obtain the pertinent data for substitution into the Goldman equation. Another criterion for electrogenic pumps has been applied to epicotyl and coleoptile cells by Higinbotham, Graves and Davis (1970), the effect of metabolic inhibitors. It is argued that if an inhibitor such as dinitrophenol (DNP) or cyanide causes a rapid depolarization of the potential then it is affecting an electrogenic component of the transport process. In the absence of an electrogenic pump the inhibitor would be expected to cause a gradual fall in the potential as the diffusion potential was slowly dissipated due to the even-ing out of the ion concentrations on each side of the membrane after the cessation of active transport. The only published experiments on the effect of metabolic inhibitors on the potential of roots have been those of Lyalin and Ktitorova (1969). They observed a rapid depolarization of the membrane potential of root cells of *Trianea bogotensis* by DNP (5×10^{-4} mol dm^{-3}). The magnitude of the effect however was not very great, the DNP causing a decline in the potential from -122 mV to approximately -110 mV. The electrogenic component of the potential thus appeared to be very small if indeed there was an electrogenic component. Ginsburg (1972) has argued that from a theoretical standpoint the use of metabolic inhibitors is not a satisfactory criterion for electrogenic pumps. Certainly it is not always easy to distinguish the effect of inhibitors on specific metabolic processes from undesirable side effects such as the alteration of membrane permeability. A much more rigorous criterion must be applied such as the use of the Goldman equation before we can be sure about the nature of the membrane potential in roots.

The Ussing—Teorell equation

The Nernst equation is applicable only to situations in which the ions involved are in flux equilibrium. In growing cells and tissues this situation seldom occurs and this limits the use the plant physiologist can make of a very useful relationship. Fortunately Ussing and Teorell developed independently in 1949 an expression which can be used in place of the Nernst equation in non-equilibrium conditions provided the fluxes of the ions across the membrane are known. The Ussing–Teorell equation relates the fluxes to the chemical activities of an ion as follows.

$$\frac{\phi_j^{oi}}{\phi_j^{io}} = \frac{C_j^o}{C_j^i \exp\left(z_j \frac{FE_M}{RT}\right)} = \frac{\bar{\mu}_j^o}{\bar{\mu}_j^i}. \tag{4.16}$$

The fluxes of the ion j from the outside to the inside of the membrane (ϕ_j^{oi}) and from the inside to the outside (ϕ_j^{io}) are determined with the use of radioactive isotopes. The techniques are described in Chapter 5. The other terms in equation (4.16) have been explained before. If the flux ratio is greater than that predicted by the right hand side of the equation then the ion is likely to be actively transported inwards. The converse is true for active efflux. The Ussing–Teorell equation is a very good test for active transport although it may not be valid under some conditions such as when the ion fluxes are not independent but are being dragged along by, for instance, a flux of water.

A derivation of equation (4.16) gives an expression for the electrochemical potential gradient on the ion, that is, the driving force on it can be calculated.

$$\bar{\mu}_j^o - \bar{\mu}_j^i = z_j F(E_M - E_N) \tag{4.17}$$

where E_M = membrane potential and E_N = Nernst potential of the ion j. Driving forces are usually expressed in J x 10^3. mol^{-1}.

5 Kinetic studies

Application of enzyme kinetics to plant roots

Osterhout (1936) was one of the first to suggest that ion selectivity of cells may be due to the association of the ions to be taken up with specific substances in the membrane. He describes a model system with an artificial membrane made of guiacol and p-cresol through which potassium is transported by association with the membrane in the following way:

$$KOH + HG \rightleftharpoons KG + H_2O \qquad (5.1)$$

The membrane-potassium complex KG diffuses through the membrane and on the inside which is assumed to be more acid than the outside,

$$KG + H_2CO_3 \rightleftharpoons KHCO_3 + HG \qquad (5.2)$$

He called KG the 'carrier' and pointed out that metabolic energy must be involved if transport across the membrane results in an increase in chemical potential.

Epstein and Hagen (1952) in an important paper, approached the uptake of alkali cations by barley roots on the basis that the ions bind with specific carriers in the membrane. This process was assumed to be analogous to the binding of substrate to an enzyme.

$$E + S \underset{K_2}{\overset{K_1}{\rightleftharpoons}} ES \tag{5.3}$$

$$ES \underset{K_4}{\overset{K_3}{\rightleftharpoons}} E + P \tag{5.4}$$

where E = enzyme, S = substrate and P = product. The transported ion is considered to be analogous with the substrate and the carrier to be analogous with the enzyme. Epstein and Hagen pointed out that the postulated carriers could in fact be enzymes but need not necessarily be so.

Assuming that transport from the inside of the membrane to the outside is negligible (i.e. $K_4 = 0$) then the velocity of the ion uptake at ion concentration S is given by

$$v = \frac{V(S)}{K_m + (S)} \tag{5.5}$$

where v = velocity of uptake reaction, V = maximum velocity of uptake at which the carrier is saturated and K_m = Michaelis constant.

Taking reciprocals on both sides the equation becomes linear (Lineweaver and Burk).

$$\frac{1}{v} = \frac{K_m}{V(S)} + \frac{1}{V} \tag{5.6}$$

On plotting $1/v$ against $1/(S)$ a straight line is obtained and V and K_m can be found (Fig. 5.1). On plotting their data for the uptake of rubidium by barley roots in this way Epstein and Hagen obtained linear curves. The presence of potassium in the medium altered the slope and increased K_m for rubidium. They concluded that competitive inhibition was occurring with rubidium and potassium competing for the

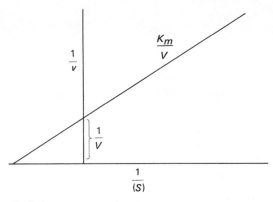

Fig. 5.1

A double reciprocal plot providing a linear
transformation of the Michaelis - Menten equation.

same carrier in the membrane. This approach was extended to
the study of chloride which was found to compete with
bromide (Epstein, 1953).

Anomalies in the results became apparent when relatively
high external ion concentrations were used. For example
Epstein and Hagen found that at rubidium concentrations
below 1 mmol dm^{-3}, sodium had no effect on the rate of
rubidium uptake, but at high external concentrations of
rubidium, sodium competed with rubidium. This and other
findings led Epstein to the conclusion that there are two sets
of binding sites for rubidium transport, one of them specific
for rubidium and potassium and the other with affinity for
sodium as well (Epstein, 1966).

Some evidence for these two patterns of binding is
illustrated in Fig. 5.2 which shows the uptake of labelled
potassium by excised barley roots over a wide range of
external KCl concentration. Two uptake phases in the form
of hyperbolae are manifest. The first system of sites appears
to be saturated at around 0.2 mmol dm^{-3} K but when the

75

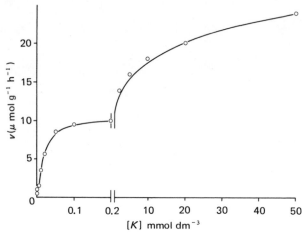

Fig. 5.2

Rate of potassium absorption by excised barley roots in
relation to the external potassium concentration. Note
the change of scale on the abscissa. After Epstein (1966).

external concentration is increased a new hyperbolic relationship is revealed which is superimposed on the first.

The characteristics of the two systems with respect to
potassium are summarized in Table 5.1. The two uptake
processes which have become known as system 1 and system
2 have been demonstrated in leaves and stems as well as
roots and in dicotyledons as well as monocotyledons. Epstein
(1973) has summarized evidence for the dual isotherms from
a total of 17 species involving uptake of both cations and
anions. Clearly they are a widespread and well documented
phenomenon.

An example of the results that have been obtained is
provided by the data of Jackman (1965). He studied the
uptake of rubidium by excised roots of barley, perennial rye
grass, mung beans and subterranean clover. Some of his
results are set out in Table 5.2. He found the two uptake
systems in all four species. However it can be seen that the

Table 5.1

Characteristics of kinetic systems 1 and 2 with respect to the uptake of potassium (From Epstein, 1966).

System 1	System 2
high affinity from K^+	low affinity for K^+
K^+ uptake unaffected by Na^+	K^+ uptake competitively inhibited by Na^+
K^+ uptake indifferent to rate of uptake of anions	Anions have great effect on K^+ uptake
Ca^{++} stimulating	Ca^{++} inhibitory
Uptake shows simple isotherm	Uptake shows complex of hyperbolic isotherms

capacities (Vmax) of the first mechanism operating below 1 mmol dm^{-3} rubidium are much greater for barley and rye grass roots than for the roots of the two dicotyledonous plants. The capacities of system 2 in all four species were, on the other hand, found to be very similar. Jackman suggests that these findings agree with the performance in the field of these species. Clovers respond well to applied potassium when grown with grasses but where deficiency conditions

Table 5.2

Km and Vmax constants for the uptake of rubidium by excised roots of barley, ryegrass, mungbean and subterranean clover. From Jackman (1965).

	System 1		System 2	
	Km mmol dm^{-3} Rb	Vmax mmol g^{-1} h^{-1}	Km	Vmax
Barley	0.017	6.6	7.1	4.2
Ryegrass	0.012	7.2	21.9	15.5
Mungbean	0.012	0.96	35.9	11.4
Subterranean clover	0.008	1.11	17.4	6.1

exist, when only system 1 is operating, the legumes appear to compete less successfully with the grasses.

The location of the two mechanisms

There has been some controversy about the location of systems 1 and 2. Do both mechanisms occur side by side in the same membrane or is system 1 at the plasmalemma and system 2 at the tonoplast? While it is generally agreed that system 1 is located in the outer membrane the site of system 2 has been a point of argument. Epstein and his colleagues (Epstein, 1972; Welch and Epstein, 1968, 1969) take the view that system 2 occurs in the plasmalemma and operates in parallel with system 1. On the other hand Laties and his co-workers have obtained evidence which suggests that system 2 is at the tonoplast. Torii and Laties (1966) in a study of rubidium uptake by corn roots failed to find system 2 in non vacuolated cells and concluded that this was because it was confined to the tonoplast which was assumed to be absent. Transport of ions across the root to the shoot does not appear to involve the vacuoles (see Chapter 7) and Lüttge and Laties (1966) found that system 1 was the only system operating in the transport of rubidium to the shoot of corn seedlings. This was offered as further evidence for the location of system 2 at the tonoplast. However, extrapolation of the concept of dual uptake mechanisms from the individual cell to transport across the whole root is questionable.

The view that system 2 is at the tonoplast necessitates the assumption that at those concentrations where system 2 comes into action (i.e. above 1 mmol dm^{-3}) mechanism 1 is not rate limiting for entry into the cytoplasm. If system 1 is always rate limiting then the rate of operation of system 2 could not exceed the rate of working of system 1. Laties and his colleagues have assumed that at concentrations above 1 mmol dm^{-3} the plasmalemma becomes more permeable to ions and they enter the cytoplasm at diffusion rates greater than the maximum uptake rate of system 1. Thus system 2 at the tonoplast becomes the overall rate limiting step. This is difficult to accept because at the high rates of uptake of system 2 it means that there must be rapid diffusion of ions into the cytoplasm and therefore there would be no effective cytoplasmic control over salt uptake. Later on in this chapter we will see that both uptake and loss of ions is under a high degree of control by the outer membrane even at high external concentrations.

Welsh and Epstein (1968, 1969) have provided indirect evidence against system 2 being located at the tonoplast but perhaps the best evidence in favour of it being at the plasmalemma has been provided by Gerson and Poole (1972). They found that chloride uptake into root tips of mung beans increases steadily with increasing external concentration from $1 - 60$ mmol dm^{-3}. They found active transport of Cl against the electrochemical potential gradient in the relatively non-vacuolate cells of the root tip (Fig. 5.3). They measured the internal chloride concentration with Cl^- specific micro-electrodes and also measured the electrical potential in order to calculate the electrochemical potential gradient for Cl^-. Careful observation of photographs of longitudinal sections of the root tip enabled them to calculate the amount of cytoplasm in the cells. They calculated that 77 per cent of the intra-cellular volume was occupied by cytoplasm, making the likelihood that the tip of the micro-electrode was in cytoplasm rather than the vacuole very high. Thus their results provide direct evidence that the second mechanism brings about active transport at the plasmalemma over and above that brought about by system 1.

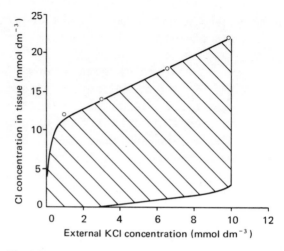

Fig. 5.3

Chloride concentrations in the tips of Mung bean roots after 21h of uptake from KCl. Uptake was assumed to be mainly in the cytoplasm as the cells were only partially vacuolated. Shaded area denotes actively accumulated Cl^-. After Gerson and Poole (1972).

Criticisms of the enzyme kinetics approach

The application of enzyme kinetics to the study of salt uptake by roots has not been without its critics. Criticism has been made both of the methods employed and of the interpretation of the dual isotherms obtained.

Higinbotham (1973) in a review article has underlined three requirements for the experiments.

(1) Low salt tissue.
(2) Short term uptake periods.
(3) Use of a wide range of external concentrations.

In order to keep the constant K_4 in equation (5.4) negligible the internal salt concentration of the tissue must be low and the period of uptake must be short to reduce the possibility of a high efflux of salt. In practice tissue starved of salt and uptake periods of less that 30 minutes are employed (Epstein, 1972b). There is a rapid reduction of uptake if longer periods are employed and after 3h, uptake of potassium by barley roots drops to about 30 per cent of its original value (Laties, 1969). Higinbotham has questioned whether this short-term rapid uptake has any relationship to the long-term steady state uptake which is obviously the more important process.

Higinbotham also draws attention to the effect changing the external potassium concentration will have on the electrical potential difference across the plasmalemma. A tenfold increase in the external potassium concentration generally reduces the transmembrane potential by $30 - 50$ mV (Etherton and Higinbotham, 1960; Lyalin and Ktitorova, 1969). This would tend to cause an increase in anion influx and a reduction in cation influx. Furthermore there is evidence that the active transport system may bring about a separation of charges at the membrane (Higinbotham, Graves and Davis, 1970) so a change in the trans-membrane potential would have an effect on its rate of working.

Pitman, Mowat and Nair (1971) have made the observation that an implicit assumption of the kinetic approach is that *only* the carrier and its relation to the concentration is limiting uptake. They suggest that this assumption is not always warranted and cite as an example their own experiments in which they find that uptake of salt by low salt barley roots is correlated with sugar level. They put forward the idea that multiple components in uptake kinetics could be explained in terms of the effect of salt on metabolism.

Dainty (1963) has pointed out that the concentration of an ion at the membrane or cell surface may bear little resemblance to that in the bulk solution because of the presence of unstirred layers. If the uptake of ions by a cell is rapid the solution at the cell surface will be depleted of ions and if diffusion into this zone from the bulk solution is slower than the rate of uptake a concentration gradient will be set up. The extent of this gradient may be reduced by stirring but it is not eliminated altogether. With good stirring the unstirred layer may be reduced to $10 - 300 \, \mu m$ thick (Nobel, 1970) but this is still an appreciable distance for molecules to diffuse. Reducing the concentration of the solution will increase the importance of the diffusion factor and it is highly likely that in the concentration range of system 1 particularly at concentrations below $0.1 \, mmol \, dm^{-3}$ the kinetics of salt uptake are governed not by the carrier but by diffusion across the unstirred layer. Polle and Jenny (1971) have obtained some experimental evidence for this. They found that rubidium uptake by barley roots from an external concentration of $0.009 \, mmol \, dm^{-3}$ was increased by stirring. However at an external concentration of $0.9 \, mmol \, dm^{-3}$ stirring had no effect.

Most experiments on uptake have been carried out with roots which have a microflora of bacteria and fungi on them. The possibility that the dual isotherms are a reflection of uptake by the micro-organisms has been put forward by Barber and Frankenburg (1971). At $0.01 \, mmol \, dm^{-3}$ the absorption of phosphate by non-sterile roots of barley was three times that of sterile roots. They suggest that uptake in the region of system 1 may be largely due to the micro-organisms rather than the root. Epstein (1972a) has convincingly rebutted this suggestion citing the widespread occurrence of the two systems in various tissues including those in which the effect of micro-organisms would be minimal. He (Epstein, 1968) and indeed Barber himself (1972) have demonstrated dual isotherms in sterile barley roots. They effect of micro-organisms in relation to the dual uptake systems, with the possible exception of phosphate uptake at very low concentrations, appears therefore to be unimportant.

The evidence for the dual isotherms is so strong that their existence is beyond question. Their significance however is another matter and several alternatives have been suggested to explain their role in salt uptake. Epstein and Rains (1965) studied the high concentration isotherm for potassium uptake by barley roots and found that it could be separated into four distinct phases, each a hyperbolic isotherm. An example of this type of curve is shown in Fig. 5.4. They interpreted this multiphasic isotherm as representing a spectrum of carrier sites differing slightly in their affinity for the ion. It seems perfectly possible to include the low concentration system 1 isotherm in this family of curves so that instead of dual isotherms we have a multiplicity of

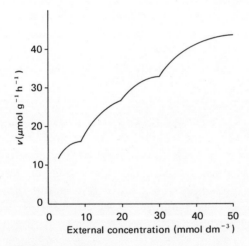

Fig. 5.4

A typical complex curve obtained when rate of uptake is plotted against external concentration in the concentration range of system 2.

isotherms. The question then arises; are these isotherms representative of a number of carrier systems or are they just facets of one uptake process?

Nissen has suggested the latter possibility on the basis of his results for sulphate uptake by barley roots and leaf slices (Nissen, 1971). Like Rains and Epstein he obtained a number of phases in the uptake of SO_4 above 10^{-6} mol dm^{-3}. Each phase obeys Michaelis-Menten kinetics and the kinetic constants increase in a fairly regular manner. He finds both system 1 and system 2 show several distinct isotherms. He has looked at the dual isotherms obtained earlier by other workers, and by replotting the data claims to show that they too show a multiplicity of isotherms (Nissen, 1973). He concludes that salt uptake can be described by a single multiphasic isotherm that is, it is a single process through the whole concentration range. He further suggests that the discontinuities in the uptake curve are due to phase changes in the membrane caused by the increase in concentration of the ions.

Another unary interpretation has been put forward by Gerson and Poole (1971). They propose that both mechanisms may be attributed to a single uncharged carrier species at the cell membrane. The ion-carrier complex would then have the same charge as the carried ion resulting in kinetics which are similar to those of dual mechanisms. For ions which move through the membrane exclusively on a carrier the concentration of the ion in the surface of the membrane C' has a hyperbolic relationship to the solution concentration C_0.

$$C' = \frac{\alpha\, C_0}{K + C_0} \tag{5.7}$$

K, they suggest is equivalent to K_m of the Michaelis-Menten equation. α is a proportionality constant.

Combining equation (5.7) into the Goldman ionic flux

equation they derive the following relationship:

$$\text{Flux in} = -\alpha \frac{zFE/RT}{1 - \exp(zFE/RT)} \frac{C_0}{K + C_0} \tag{5.8}$$

Substituting data taken from the literature into equation (5.8) they obtained influx curves resembling dual mechanisms. Gerson and Poole admit that their model is inadequate to explain all the aspects of dual mechanisms but it does underline the possible importance of the membrane potential on carrier mediated transport, an aspect which has hitherto been ignored by Epstein and his co-workers.

Thellier (1970) and Thellier, Thoiron and Thoiron (1971) have employed an approach to ion uptake kinetics based on non-equilibrium thermodynamics, which takes into account both the chemical and the electrical aspects of the process. The overall uptake process is:

$$S \rightleftharpoons P \tag{5.9}$$

where S is the ion outside the cell and P the ion inside. Considering the velocity of the uptake process v as equivalent to an electrical intensity I

$$v = I$$

With a process obeying Ohm's law

$$I = \frac{\Delta E}{r} \tag{5.10}$$

If $\Delta E = 2.3\ A \log B \dfrac{[S]}{[P]}$.

where

$$A = \frac{RT}{zF}$$

85

and B = a constant characteristic of the dynamic state of the cell;

then

$$v = 2.3 \frac{A}{r} \log B \frac{[S]}{[P]}. \tag{5.11}$$

However, the uptake process may not be ohmic and the law of varistant semi-conductors may apply.

$$I = \frac{\Delta E}{r} + (\lambda \Delta E)^m \tag{5.12}$$

where r, λ and m are characteristics of the cell structures involved in the uptake process.

Then

$$v = 2.3 \frac{A}{r} \log B \frac{[S]}{[P]} + 2.3(\lambda A) \log \left(B \frac{[S]}{[P]} \right)^m. \tag{5.13}$$

Thellier and his colleagues find that as long as $[S]$ remains sufficiently small to permit $B \frac{[S]}{[P]}$ to be much smaller than 10 the second term of equation (5.13) remains negligible. In the ohmic phase the relationship between $[S]$ and v is hyperbolic and tends to a saturation plateau. As $[S]$ rises it is the second term in equation (5.13) which becomes predominant (the non-ohmic phase) and this results in a second curve typical of Epstein's system 2.

The approaches of Gerson and Poole and Thellier and his co-workers suggest therefore that the 'dual carriers' may be merely the consequence of the properties of one uptake system. They are a promising start in trying to bridge the gulf between the enzyme kinetics and the electro-physiological approaches to this problem. One aspect of this gulf is the difference of opinion that has developed over the nature of ion movement across the membrane. The protagonists of the enzyme kinetics approach, notably Epstein, tend to the view that all ion movement across the membrane, both into and out of the cell, is mediated by carriers. On the other hand, those who envisage active ion transport driven by ion pumps assume that passive diffusion of ions goes on across the membrane at the same time although usually in the opposite direction to the active transport. A dynamic equilibrium is thus assumed to eventually develop with the influx of a given ion being balanced by an equivalent efflux. Epstein does not accept this but takes the view that the carrier mediated uptake mechanisms are poised in favour of the inward reaction with little or no backward transport from the cell to the outside. He thus envisages a static equilibrium. He puts forward some evidence for this by describing experiments in which tissues were loaded with labelled ions and then washed in the same concentrations of stable isotope. No labelled ion appears to have been lost from the tissue (Epstein, 1973). He suggests that leakage of isotope observed in similar experiments by other workers may be due to a number of artifacts as a result of the experimental procedure. However a large body of evidence has been obtained which indicates that isotopic exchange does readily occur and although some of the experimental techniques may be criticized most of the results cannot be dismissed in this way.

Measurement of tracer fluxes

Chemical analysis of root tissue provides a reasonably accurate measure of the ionic concentration in the vacuole as it occupies about 90 per cent of the cell. The determination of the ionic concentration in the cytoplasm has proved to be much more difficult since it occupies only about 3 per cent of the cell volume as a layer approximately 0.5 μm thick lining the cell wall (Macklon and Higinbotham, 1970).

The only direct measurements of ion activity in the

cytoplasm of roots have been made by Etherton (1968) and Gerson and Poole (1972) using ion specific micro-electrodes. Etherton found that the K^+ activity in the cytoplasm and the vacuole of pea root cells was 43 and 122 mmol dm^{-3} respectively. The reliability of the cytoplasmic value however must be in considerable doubt as the thickness of the cytoplasm is close to the limits of resolution of the light microscope and so it is difficult to see how Etherton could be sure the tip of his micro-electrode was in the cytoplasm. Gerson and Poole measured Cl^- activity in the cytoplasm of cells at the root tip of mung bean. These cells were only partially vacuolated. The calculated that in the first milli-metre of root, vacuolar volume/tissue volume was only 21 per cent. They therefore assumed that the electrode when in the cell was more likely to be in the cytoplasm than the vacuole. They found a mean Cl^- activity of 21.8 mmol dm^{-3} which was not significantly different from the value in the vacuolar sap of more mature cells further back from the root tip.

Another approach to this problem has been the study of the pattern of radioisotope efflux from tissues previously loaded with labelled ions. MacRobbie and Dainty (1958) found that tracer efflux from the large celled alga *Nitellopsis obtusa* revealed three kinetic phases, a rapid diffusional phase, which they identified with the free space of the cell wall, a slower phase assumed to be loss from the cytoplasm and a third phase, the slowest of the three which they assumed to be loss from the vacuole. Analysis of the data enabled the isotope fluxes in and out of the cytoplasm and vacuole to be worked out.

Pitman (1963) and Cram (1968) have applied this tech-nique to beetroot and carrot storage tissue respectively, with considerable success. Even in these more complex tissues the three phase washout curves could be obtained. The grounds for assigning the respective phases to the cell wall cytoplasm and vacuole are not so strong as for the much simpler filaments of *Nitellopsis*, nevertheless, Cram advances some reasonable evidence for the validity of this interpretation.

If a cell containing no radioactive ions is placed in a solution containing radioisotope then the initial rate of entry of labelled ions is given by

$$\frac{dC_i^*}{dt} = J_{in} \frac{AC_0^*}{VC_0} \tag{5.14}$$

where A = surface area of tissue, V = tissue volume, C_1^* = internal concentration of labelled ions, C_0 = total external ion concentration, C_0^* = external concentration of labelled ions and J_{in} = influx of the ion.

After some time, when radioactive ions are crossing the cell boundary in both directions the rate of change of internal activity is given by

$$\frac{dC_i^*}{dt} = J_{in} \frac{AC_0^*}{VC_0} - J_{out} \frac{AC_i^*}{VC_i} \tag{5.15}$$

where C_i = total internal ion concentration and J_{out} = efflux of the ion.

In an isotope washout experiment the labelled ion is prevented from accumulating in the external solution by frequent replacement of the external solution with fresh solution of stable isotope at the appropriate concentration. In this situation C_0^* can be taken as zero and the solution of equation (5.15) becomes

$$C_i^* = C_{init}^* \exp\left(-\frac{J_{out}A}{C_iV} t\right) \tag{5.16}$$

where C_{init}^* = the initial concentration of labelled ion in the cell. If $\ln C_i^*$ is plotted against time the curve will be a straight line of intercept C_{init}^* and slope (J out A/VC_i).

89

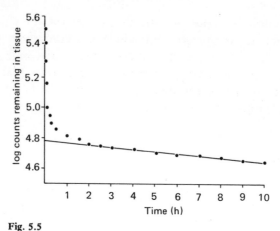

Fig. 5.5

Efflux of ^{22}Na from intact roots of sunflower into 12.2 mmol dm^{-3} NaCl. Data obtained in the author's laboratory by Una H. Shepherd.

In experiments with root tissues the slope of the washout curve is termed k, the rate constant for efflux as the surface area and the volume are difficult, if not impossible, to determine. In practice many workers who have used the washout technique have published curves in which $\log_{10} C_i^*$ is plotted against time. The necessary adjustment to the value of k is made in later calculations.

A typical efflux curve is shown in Fig. 5.5. This was obtained from a washout experiment carried out in the author's laboratory in which 0.2 g of roots of sunflower (*Helianthus annuus*) were loaded with ^{22}Na for 12h (t_{uptake}) and then washed in 12.2 mmol dm^{-3} NaCl for a further 10h. Three phases can be seen in this curve and the intercept at $t = 0$ for the slowest phase gives I_v the apparent initial content of the vacuole. The vacuolar content is subtracted from the total activity at each time interval in Fig. 5.5 and a new curve obtained by plotting the activity remaining against time. This typically shows two distinct phases (Fig. 5.6). Extrapolation

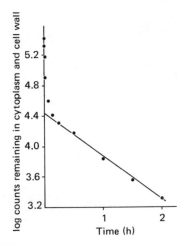

Fig. 5.6

Efflux of ^{22}Na from the cytoplasm and cell wall phases of intact sunflower roots. Data obtained by Una H. Shepherd.

of the slower phase to $t = 0$ gives I_c the apparent content of the cytoplasm. The cytoplasmic component may then be subtracted from the total activity in Fig. 5.6 and a third curve obtained by plotting the activity remaining against time (Fig. 5.7). The two phases in Fig. 5.7 represent the contents of the cell wall and the surface film; The intercepts at $t = 0$ give the total contents of these phases I_{cw} and I_s. The rate constants k_v, k_c, k_s and k_{cw} can be calculated from the respective curves.

The fluxes and ionic contents of the various compartments obtained from the curves are only apparent values as they do not take into account the effect of concurrent opposing fluxes. For example, the cytoplasmic content (I_c) is an underestimate because it does not take into consideration the transfer of ions between vacuole and cytoplasm, only the appearance of tracer in the outside solution.

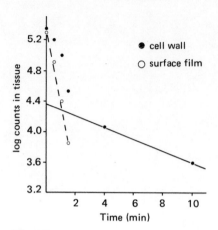

Fig. 5.7

Efflux of ^{22}Na from the cell wall and surface film of sunflower roots. ($\circ$ —— $\circ$); surface film. ($\bullet$ —— $\bullet$); efflux from cell wall. Data obtained by Una H. Shepherd.

The true fluxes are as follows:
Flux from outside the cell into the cytoplasm

$$J_{oc} = k_c I_c + \frac{I_v}{t_{uptake}}. \tag{5.17}$$

Flux from the vacuole to the cytoplasm

$$J_{vc} = \frac{J_{oc} k_v Q_v}{k_c I_c} \tag{5.18}$$

Flux from the cytoplasm to the outside solution

$$J_{co} = k_c I_c + k_v Q_v. \tag{5.19}$$

Flux from the cytoplasm to the vacuole

$$J_{cv} = J_{oc} + J_{vc} - J_{co}. \tag{5.20}$$

Q_v = the amount of ion in the vacuole estimated by chemical analysis of a parallel sample of tissue at the beginning of the washout period minus the apparent contents of the cytoplasm and cell wall.

An approximate cytoplasmic content is given by:

$$Q_c = I_c \frac{J_{cv} + J_{co}}{J_{oc}} \cdot \frac{J_{cv} + J_{co}}{J_{co}} \qquad (5.21)$$

or more simply

$$Q_c = \frac{J_{cv} + J_{co}}{K_c} . \qquad (5.22)$$

This in turn can be used to calculate the true vacuolar content $Q_{v'}$

$$Q_{v'} = Q_v + I_c - Q_c . \qquad (5.23)$$

$Q_{v'}$ can then be used to recalculate the fluxes to give closer approximations. Q_c can then be recalculated from the new flux values. If the value obtained is markedly different from the original value, the cycle of calculations is repeated until a stable figure, the true cytoplasmic content $Q_{c'}$ is obtained. Further details of the theory of this technique are given in the papers by Pitman (1963) and Cram (1968).

The concentration of ion in the cytoplasm can be calculated from an estimate of the cytoplasmic volume. If the transmembrane potential is known, this value, together with the fluxes and concentrations can be substituted into the Ussing-Teorell equation (Chapter 4). This will indicate if active transport of the ions is occurring across the plasmalemma or the tonoplast and also show the direction of such transport.

Permeability coefficients

Data obtained from washout experiments enable permeability coefficients to be calculated from relationships derived from the Goldman equation.

93

$$\text{e.g.} \quad J = -P \frac{zFE/RT}{1 - \exp zFE/RT} [C_o - C_i \exp zFE/RT]$$

$$(5.24)$$

where J = net flux of the ion, P = permeability coefficient, C_o = external ion concentration and C_i = internal ion concentration. Equation (5.24) is only applicable for passive transport and cannot be used where active transport of an ion is involved. Where an ion appears to be moving with the electrochemical gradient across a particular membrane it is generally felt that use of equation (5.24) is valid even though the ion may be actively transported in the opposite direction in that membrane. For instance if an ion was found to be moving across the plasmalemma in to the cell against the electrochemical gradient its flux in the opposite direction may be used to calculate P thus:

$$J_{co} = P \frac{zFE}{RT} \frac{C_i \exp zFE/RT}{1 - \exp zFE/RT}. \qquad (5.25)$$

Some criticisms of the interpretation of isotope flux data

Washout experiments are virtually the only way we can measure many of the important parameters involved in salt transport across the two membranes of the plant cell. Important conclusions such as the location of different active transport systems on the plasmalemma and the tonoplast depend upon the validity and soundness of this technique. It is therefore important to bear in mind its shortcomings which even its major exponents have to their credit, underlined in their papers.

A basic assumption of the method is that the three main components of the efflux curve can be identified with the three main compartments of the cell, namely the cell wall, the cytoplasm and the vacuole and that they are connected in series. However in a complex tissue like the root it is possible that phases in the washout curve could be due to say different layers of cells in the tissue. Furthermore, movement of isotope from cell to cell by way of the plasmodesmata could have a serious effect on the size of the cytoplasmic component. The series model may not hold even for single cells as there is some evidence for two parallel kinetic compartments in the cytoplasm from the work of MacRobbie (1969) on the algae *Nitella* and *Tolypella*. Her results suggest that ions can traverse the cytoplasm and enter the vacuole in discrete 'quanta' without mixing with the rest of the ions in the cytoplasm. This means that there may be a direct connection between the vacuole and the outside solution, by-passing the main bulk of the cytoplasm. A more complex model may therefore be required to explain the pattern of efflux in some washout experiments.

Macklon and Higinbotham (1970) found that in a typical washout experiment the amount of labelled potassium removed from pea epicotyl tissue loaded for 12h was about 3.7 per cent of that taken up. Of this amount the loss from the surface film and cell walls amounted to 57 per cent therefore only about 1.6 per cent of the isotope taken up by the cytoplasm and vacuole was subsequently washed out. Small errors in measurement of the slopes and intercepts of the efflux curves will therefore lead to very large errors in the calculated amounts of ion in the cytoplasm. Furthermore the volume of the cytoplasm is very difficult to measure and estimates can vary between 0.5 and 3.4 per cent of the volume of a fully vacuolated cell (Macklon and Higinbotham, 1970). Such a wide variation could result in a seven-fold error in the cytoplasmic concentration.

Most washout experiments on roots have been carried out with excised tissue and the effect of isotope loss from the cut ends has often been ignored. Pallaghy and Scott (1969) in endeavouring to obtain reasonably homogeneous samples of broad bean roots, removed the stele from 10 mm lengths of tissue. One wonders about the effect of such drastic treatment on the behaviour of the tissue. They studied fluxes in this material for up to five days and report that oxygen consumption fell by 60 per cent in this time, which is perhaps not surprising.

A most serious omission has been the failure to allow for the exudation from the cut xylem vessels of excised roots and epicotyls. Pitman (1971) found that about 75 per cent of labelled chloride diffusing out of excised barley roots came from the stele. This is a very large proportion and must throw considerable doubt on the validity of experiments of this kind in which exudation has been ignored.

Despite this list of possible disadvantages the isotope efflux technique has provided us with some data about the fluxes in various compartments in the cell and some figures for ionic concentrations in the cytoplasm, which hopefully bear some relation to the true situation. Whilst this approach has been applied recently to epicotyl and coleoptile tissues by Macklon and Higinbotham (1970) relatively few studies have been made on roots. Data for young growing roots obtained by various laboratories are given in Table 5.3. It can be seen firstly that the data, especially for permeability coefficients, are scanty and secondly, that cytoplasmic concentrations bear little relationship to the vacuolar concentrations. In some plants the cytoplasmic concentration is higher than that of the vacuole but in other species it is the other way round. These differences may reflect true species differences or more likely are due to inaccuracies in the data. Obviously much more information of this kind is needed before generalizations can be made about compartmental concentrations and membrane permeabilities.

In a study of the compartmentation and fluxes of sodium, potassium and chloride in oat coleoptile cells, Pierce and Higinbotham (1970) found that at the plasmalemma, accord-

Ion concentrations (m mol dm^{-3}) in the cytoplasm and vacuole and permeability coefficients (cm s^{-1} × 10^{-8}) for the plasmalemma of young absorbing roots.

Tissue	Ion	External concentration	Cytoplasmic concentration	Vacuolar concentration	P	Reference
Mungbean	Cl$^-$	10	27.1	24.7		[‡]Gerson and Poole, 1972
Pisum sativum	K$^+$	10	43	122		[‡]Etherton, 1968
Barley	Na$^+$	7.5	70*	29†		Pitman and Saddler, 1967
Barley	K$^+$	2.5	102*	74†		Pitman and Saddler, 1967
Vicia faba	K$^+$	1.0	20*	33.5†		Pallaghy and Scott, 1968
Triglochin	K$^+$	2.0	32.3	48.5	0.016	Jeffries, 1973
maritima	K$^+$	8.0	70.8	11.6	0.018	Jeffries, 1973
T. maritima	Na$^+$	1.0	74.8	75.2	0.071	Jeffries, 1973
T. maritima	Na$^+$	100	110.1	81.1	0.013	Jeffries, 1973
Vicia faba	K$^+$	1.0			22	Scott, Gulline and Pallaghy, 1968
Vicia faba	Na$^+$	1.0			6	Scott, Gulline and Pallaghy, 1968
Vicia faba	Cl$^-$	3.0			0.043	Scott, Gulline and Pallaghy, 1968
Zea mays	K$^+$	0.2	21.0	21.5		Lüttge and Laties, 1967
Onion	K$^+$	1.0	100	83	0.31	Macklon, 1975a
Onion	Na$^+$	1.0	9.4	44	0.92	Macklon, 1975a
Onion	Cl$^-$	1.0	30	24	0.052	Macklon, 1975a
Onion	Ca^{++}	1.0	10.2	1.4	3.7	Macklon, 1975b

*Calculated from values of Qc assuming that the cytoplasm occupies 5% of the total fresh weight of the tissue.
†Calculated from Qv assuming vacuole occupies 85% of the fresh weight.
‡Measured with micro-electrodes.

ing to the Ussing-Teorell equation, Na^+ is actively pumped out and Cl^- actively transported inwards. At the tonoplast their results indicated that Na^+ is pumped in with Cl^- probably in equilibrium. They were uncertain about the behaviour of K^+. The work of Jeffries (1973) on *Triglochin maritima* and Macklon (1975) on onion, provide virtually the only complete flux analyses for young roots. Jeffries studied the status of sodium and potassium in *Triglochin* which is a halophyte and therefore likely to behave differently from the common glycophytes. He studied the behaviour of potassium over the concentration range $0.2-8$ mmol dm^{-3}. Transport of K^+ across the plasmalemma appeared to be passive at all the external concentrations although close agreement of the ratios of electrochemical potential and the flux ratios did not occur at high external concentrations. This suggests that at high external concentrations potassium may be actively pumped out. A similar situation was found for the tonoplast so that on the whole the behaviour of potassium was uncertain, as Pierce and Higinbotham also found.

A more clear cut result was obtained for the behaviour of sodium however. Some of Jeffries data are set out on Fig. 5.8. At external concentrations of 100 mmol dm^{-3} and above the ratios of the electropotentials were found to be much greater than the flux ratios across the plasmalemma, clearly indicating that sodium was being actively extruded from the cell. The ratios for the tonoplast were very similar to each other indicating that here sodium remained in equilibrium. Thus *Triglochin* behaves like the brackish and marine algae in its ability to control its internal sodium levels (Gutknecht and Dainty, 1968).

Macklon (1975a,b) investigated the behaviour of Na^+, K^+, Ca^{++} and Cl^- in onion root segments either 2 or 4 cm in length. Loss of isotope from the cut ends of the segment could be separated from efflux from the middle part of the segment by using a vessel with three compartments (Fig. 5.9).

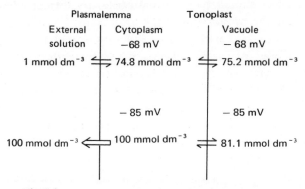

Fig. 5.8

The distribution and behaviour of sodium in the root cells of *Triglochin maritima* based on the data of Jeffries (1973). The bold arrow represents active transport and the half arrows represent passive equilibrium. The electrical potential differences were measured between the vacuole and the external solution and it is assumed that the PD across the tonoplast is zero.

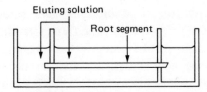

Fig. 5.9

Chamber used by Macklon (1975a,b) for following the efflux of isotope from the intact tissues and the cut ends of onion root segments.

This vessel was similar to one used earlier by Pitman (1971) to determine the proportion of the total isotope exuded from the xylem of excised barley roots. Macklon found that failure to take account of the leakage from the cut ends resulted in an over-estimate of the ion fluxes across the tonoplast although it did not alter the conclusions qualitatively. Separate measurements of ions leaking from the ends of the segments revealed that Na^1 was transported almost exclusively in an acropetal direction in the stele. On the other hand, K^+ was transported mainly in a basipetal direction and the transport of Cl^- in the stele showed no marked polarity.

The concentrations of the four ions in the cytoplasm and the vacuole computed from the isotope flux analysis are shown in Table 5.3. Substitution of the concentrations and fluxes in to the Ussing-Teorell equation (4.16) led Macklon to the following conclusions. K^+, Na^+ and Cl^- are actively transported into the cell across the plasmalemma, whilst Ca^{++} enters passively and appears to be pumped back across the plasmalemma to the outside solution. At the tonoplast, K^+ and Cl^- are in equilibrium whilst Na^+ is actively transported into, and Ca^{++} actively transported out of, the vacuole. These conclusions are summarized diagrammatically in Fig. 5.10.

Macklon's results indicate that the plasmalemma is the location of pumps for all four ions and emphasize its importance as the main site for active transport into the cell. Perhaps the most significant finding is that sodium is actively transported into the cell across this membrane as well as across the tonoplast. This supports the evidence obtained by Bowling and Ansari (1971) for a sodium influx pump in sunflower roots and by Shepherd and Bowling (1973) in the roots of several aquatic species. As potassium and sodium are actively transported across the plasmalemma in the same direction it is difficult to see how a Na-K exchange pump could be operating as suggested by some workers (see Chapter 4). This information, on the contrary, augments the

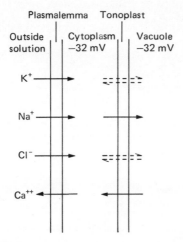

Fig. 5.10

Active and passive transport of various ions
in the plasmalemma and tonoplast of onion
root cells. Solid arrows; active transport,
broken arrows; passive movement. Based on
data of Macklon (1975a,b).

body of evidence which indicates that higher plant cells have
active transport mechanisms which are quite distinct from
those found in animal cells and in the cells of some algae.

The results from the halophyte *Triglochin* and onion, a
glycophyte, are quite different. It seems rather surprising that
Triglochin which is obviously tolerant of high sodium
concentrations, pumps Na^+ out of the cells whilst onion,
which is presumably not so tolerant of high salinities,
actually pumps it in. Obviously we require more information
of this kind from a wide variety of roots before we can hope
to understand how different species cope with the salinity
problem and how they regulate the level of other ions as well
as sodium in the cytoplasm and the vacuole.

6 Mechanisms and hypotheses

General considerations

It is true to say that we know very little about the mechanism of active transport and the details of how cellular energy is brought to bear on ions in order to transport them in to the cell. We have very little information about what goes on at the heart of the active uptake process and so we are forced to speculate and hypothesise about what might occur. It is generally agreed that the carriers inferred from kinetic analyses and the pumps suggested by the results of electro-physiological experiments, whatever their true nature, reside in the cell membranes. Active transport is therefore associ-ated with membranes and some have gone as far as to suggest that it is a characteristic only of membranes, so that if there are no membranes there is no active transport. On this argument plasmodesmata, for example, would not be ex-pected to transfer substances between cells by active trans-port as there does not appear to be an intervening membrane where the necessary driving force could be exerted. To link active transport and membranes so closely is perhaps an extreme view considering our present lack of knowledge but it does underline the undoubted importance of the bounding membranes in the active transport of ions into the cell.

Unfortunately the anatomical structure of the cell membrane has revealed little about the transport mechanism which must reside in it. Summing up a Royal Society discussion on the electron microscopy and composition of biological membranes Sir John Randall wrote 'Other than in a very broad sense, structural studies have so far given only limited guidance to possible transport mechanisms' (Randall, 1974). What knowledge we have of membrane structure comes mainly from studies on animal cells. In plants, structural studies on the membranes of root cells have been relatively few. Furthermore, these investigations have usually been aimed at membrane based processes other than salt transport so that we have to depend on other approaches for knowledge of membrane transport in the root.

Although active transport was first discovered in plant cells much of the research in recent years has been made by workers using animal cells. The result of this has been that plant physiologists have tended to extrapolate hypotheses developed for animal cells to the situation in plants and to roots in particular. This approach has however met with only limited success. It is becoming clear that the reason for this is that plant cells possess distinct transport systems which are rather different from those found in animal cells.

Before any reasonable hypothesis to explain the mechanism of active transport can be framed, some basic information about plant cells needs to be taken into account. In particular we need to consider:

(a) That cells are selective in the ions they accumulate.

(b) The biphasic and multiphasic isotherms often observed.

(c) The relationship between salt uptake and respiration.

(d) The electrical potential difference across the plasmalemma and possibly also across the tonoplast.

(e) That anion uptake is generally by active transport whereas the nature of cation uptake is in doubt.

(f) That pH has a marked effect on salt uptake.

In connection with the last point, the evidence for the pH of the cytoplasm being closely regulated by a hydrogen extrusion pump at the plasmalemma was discussed in Chapter 4. It is generally supposed that this proton pump keeps the pH of the cytoplasm at pH 6.0–7.0. If the pH of the medium surrounding the cell is outside this range then we have to take into consideration the strong possibility of there being a pH gradient across the plasmalemma.

Proton and electron fluxes

One of the earliest theories to explain the mechanism of salt uptake by plants was the electrochemical theory of Lundegårdh (1939). It is specially singled out here from a number of theories current at that time because it brilliantly anticipated some later findings and as we shall see, it is the forerunner of an important modern theory. Lundegårdh and Burström (1933, 1935) had shown that salt uptake was closely related to a component of respiration, salt respiration, and both anion uptake and salt respiration were inhibited by cyanide (Table 3.2). Lundegårdh realizes that as the terminal member of the cytochrome system is inhibited by cyanide, salt uptake could be related to the flow of electrons in the cytochromes during respiration. He postulated that the electrical potential gradient between the cell membrane and the outside solution would cause a movement of electrons from the inside to the outside of the membrane. Anions were assumed to traverse the membrane in the opposite direction in exchange.

In a later paper (Lundegårdh, 1945) he envisaged the iron atom in the cytochrome molecule as the seat of this

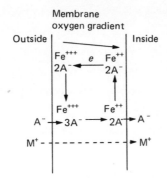

Fig. 6.1

Hypothetical scheme relating anion uptake to
electron flow in the membrane. After
Lundegårdh (1945).

electron flow according to the scheme shown in Fig. 6.1.
The general respiration of the cell was assumed to reduce the
partial pressure of the cytoplasm oxygen below its ambient
level, so giving rise to a gradient of redox potential across the
membrane. In the outer part of the membrane, which is
considered to be under oxidizing conditions, the iron in the
cytoplasm is trivalent. When the cytochrome moves, by
thermal agitation, into the reducing zone at the inner part of
the membrane, the trivalent iron is reduced to the divalent
form by acquisition of an electron passed on from the
respiratory electron chain. On moving back to the outer part
of the membrane the iron of the cytochrome is re-oxidized,
the electron released uniting with a proton, both being lost
from the cell as water. The outward flow of electrons in this
system is balanced by a counterflow of anions into the cell.
Lundegårdh realized that as the cytoplasm of the cell
appeared to be electrically negative in relation to the outside
then the work involved in active transport must be applied to
the anions, with the cations moving in to maintain the
balance between anions and cations.

Since

$$C_6H_{12}O_6 + 6H_2O + 6O_2 \longrightarrow 6CO_2 + 12H_2O$$

$$(6.1)$$

It follows that each $6O_2$ accepts $24H^+$ and $24e$ so that if one anion enters for each electron moving across the membrane then 24 monovalent anions and 24 monovalent cations would be accumulated.

i.e. $\dfrac{\text{mol salt absorbed}}{\text{mol oxygen absorbed}} = 4.$ (6.2)

For the uptake of anions according to the Lundegårdh hypothesis therefore, the above ratio would be expected. However Handley and Overstreet (1955) found ratios considerably higher than 4 for salt uptake in excised barley roots. Moreover, soon after Lundegårdh formulated his hypothesis it was found that the cytochrome system was not in the cell membrane but on the mitochondria (Du Buy, Woods and Lackey, 1950). These two blows to the hypothesis were accompanied by another equally serious setback. 2,4-Dinitrophenol (DNP) was found to uncouple phosphorylation from oxidative respiration in carrot tissue, bringing about a reduction in salt uptake whilst the oxygen uptake increased (Robertson, Wilkins and Weeks, 1951). As Robertson has pointed out, this may either mean that salt uptake is dependent on ATP formation or that both active transport and ATP formation depend on a common process which is inhibited by DNP (Robertson, 1968).

The Lundegårdh hypothesis appeared to be disproved in the early 1950s but it had some basic ingredients of a later theory that has proved to be more successful. They are: an electron flow in the membrane, an efflux of protons and a potential difference across the membrane. Also it is pertinent to point out that anion uptake in plant cells has without exception been found to be by active transport whilst the nature of cation transport is still in doubt.

The chemi-osmotic theory

It seems a reasonable assumption that the energy required for active transport comes from ATP. As DNP inhibits both ATP formation and salt uptake, knowledge of how ATP is formed may reveal the mechanism of salt uptake. However the connection between the oxidative reactions of respiration and ATP synthesis which DNP uncouples is obscure. Mitchell has suggested a scheme which attempts to link the two processes in the mitochondrial membrane. His ideas have been modified a number of times since he first put forward his hypothesis in 1961, but basically he envisages a separation of charge occurring across the mitochondrial membrane due to electron flow through the cytochrome system, very much along the lines proposed earlier by Lundegårdh (Mitchell, 1961, 1966, 1970, 1973).

The mechanism of the proposed charge separation is shown in Fig. 6.2. A hydrogen carrier takes hydrogen to the

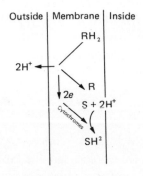

Fig. 6.2

A mechanism for charge separation in the mitochondrial membrane. After Mitchell (1970).

outer side of the membrane where it is oxidized, releasing a proton which is lost to the outside and an electron which is transferred through the cytochrome chain eventually to reduce another hydrogen carrier. Several of these systems linked in series in the membrane will result in a flow of protons from the inside to the outside (from right to left in Fig. 6.2). A pH gradient is thus set up across the membrane with the outside more acid than the inside.

Mitchell's new contribution was to suggest that an ATPase enzyme in the membrane could be affected by the charge separation across the membrane. ATPase catalyses the breakdown of ATP to ADP according to the following reaction

$$ATP^{4-} + H_2O \longrightarrow ADP^{3-} + HPO_4^{2-} + H^+. \qquad (6.3)$$

Hydrolysis of ATP to ADP inside the mitochondron has been observed to be accompanied by an extrusion of two hydrogen ions to the outside (Mitchell and Moyle, 1965). Mitchell suggested that extrusion of protons during oxidation would slow down ATP hydrolysis by the membrane ATPase and should even reverse the reaction to bring about ATP synthesis if the pH outside the membrane was low enough. As charge separation necessarily results in a potential difference across the membrane (interior of the mitochondrion negative) artificially increasing this electrical potential would also cause synthesis of ATP. Mitchell calculated that a pH gradient of 3.5 or a potential of 210 mV would keep the reaction in equilibrium.

Mitchell developed the chemi-osmotic hypothesis, as it is called, with the energy exchanges of the mitochondrion in mind but some of the best evidence for the hypothesis has come from the work of Jagendorf and his colleagues on chloroplasts. The chloroplast lamellae under normal circumstances have a pH gradient which is opposite in direction to that of the mitochondrial membrane, i.e. the inside of the lamella is relatively acid and the outside relatively alkaline.

When this pH gradient was artificially enhanced, strikingly large amounts of ATP were synthesized even in the dark. It thus appeared that the pH gradient developed during photosynthesis was responsible for the synthesis of ATP. DNP and other uncouplers of phosphorylation were found to inhibit this ATP synthesis (Hind and Jagendorf, 1965; Jagendorf and Uribe, 1966). The uncoupling of the ATP synthesis by inhibitors of this type is now generally believed to be due to their effects on the membrane, which is thought to become leaky causing the pH gradient to be dissipated. Evidence that the mitochondrion can also synthesize ATP in response to a trans-membrane pH gradient was more difficult to obtain. After a number of workers had tried and failed Reid, Moyle and Mitchell (1966) observed mitochondrial ATP synthesis when the external medium was changed from pH 9.0 to 4.3.

The chemi-osmotic hypothesis therefore provides a link between the oxidative reactions and ATP synthesis in the mitochondrion. As ATP synthesis is an energy requiring process some workers have postulated that a high energy intermediate is involved. However, no such substance has been found and the high energy intermediate appears to be not a compound but the proton gradient across the mitochondrial membrane. There is increasing evidence for the operation of this type of energy interchange in the membranes of mitochondria, chloroplasts and also bacterial cell membranes. Although the details of the process are still not clear the important point is that charge separation occurs providing a supply of potential energy which can be utilized in various processes including active transport of ions across the membrane. Mitochondria have been found to accumulate ions, particularly divalent cations like calcium and it is probable that these enter in response to the electrical and hydrogen ion gradients developed across the mitochondrial membrane.

Is the chemi-osmotic hypothesis relevant to ion uptake by

the cells of the root? Is it possible that charge separation resulting from ATP hydrolysis is the basis of active salt transport across the plasmalemma? Certainly most of the ingredients of the Mitchell scheme appear to exist in the plasmalemma of root cells. There is:

(a) An electrical potential difference which is affected by inhibitors like DNP and cyanide which suggests that there might be a metabolic charge separation across the membrane.

(b) The likelihood of a pH gradient as already discussed.

(c) ATPase in the membrane. Evidence for this is discussed in the next section.

There is however a vital ingredient which is missing, namely the electron carriers. Lundegårdh's hypothesis foundered because of the location of the cytochrome system in the mitochondrion and not in the plasmalemma. Are electron carriers in the plasmalemma required if chemiosmosis is to work, and if so are they likely to be similar or different from those in the membranes of the mitochondria and chloroplasts? Obviously we need to investigate the plasmalemma to find out if it contains substances which can act as electron carriers. We also require to improve our techniques in order to measure pH gradients across the plasmalemma and to try to correlate them with ATP turnover and ion fluxes.

Cation stimulated ATPases

In animal cells a close relationship has been established between active cation transport and membrane bound ATPase activity. In a number of animal tissues there is a close coupling between sodium and potassium transport with the sodium usually extruded from the cell and potassium accumulated. The kinetics of this coupled Na-K transport have been found to closely resemble the kinetics of the

ATPase present in the membrane. This transport ATPase is stimulated by sodium and potassium presented together and also requires the presence of magnesium for maximum activity. It is inhibited by calcium and by the cardiac glycoside ouabain which also inhibits coupled Na-K transport. Considerable work has been carried out on this system because of its obvious importance and for further information the reader should refer to the review by Whittam and Wheeler (1970).

Attempts have been made to extend this experimental approach to the plant cell and several reports of cation stimulated ATPases in roots have been published (e.g. Dodds and Ellis, 1966; Greuner and Neumann, 1966; Fisher and Hodges, 1969). The use of histochemical methods has shown them to be present in the membranes (Poux, 1967; Hall, 1969). The enzymes isolated from roots and other plant tissues differ in several ways from those found in animal cells. In animals the ATPases exhibit synergism, that is the stimulation in activity produced by sodium and potassium given together is greater than their total effect when given separately. In plant cells no synergism has been observed except for an isolated report by Kylin and Gee (1970). Ouabain, with one or two exceptions, has been found to have no inhibiting effect on the activity of plant ATPases. The occurrence of ouabain sensitive Na-K exchange has been reported for some algae but not others (MacRobbie, 1970; Raven, 1971). In the higher plant the evidence for coupled Na-K transport is not conclusive (see Chapter 4) so it is perhaps not surprising the synergism and inhibition by ouabain are not characteristics possessed by ATPases from higher plants.

Unlike the work with animal cells there has been a relative lack of success in attempts to correlate membrane ATPase activity with cation uptake by roots. One of the reasons for this is because a useful specific inhibitor of plant ATPase activity which also inhibits salt uptake, comparable with the role of ouabain in animal cells is lacking. Another reason may be because plant cells, not being as specialized as some animal cells such as erythrocytes and nerve cells, contain more enzyme systems and in particular more types of ATPase than animal cells, therefore making them more difficult to study. Hall (1971) was able to separate several ATPases from wall preparations of barley roots by disc electrophoresis and he pointed out that if only one of them was associated with transport its effect would be masked by the presence of the other forms when they were present together in the cell or in a crude extract.

A third reason for the lack of evidence of correlation between ATPase activity and cation uptake may be that specific cation transporting ATPases do not occur, bearing in mind that the evidence for active cation uptake in plant cells in inconclusive. On the other hand, anion stimulated ATPases seem a greater possibility as active anion transport is the norm in plant cells in contrast to animal cells. In this connection Ratner and Jacoby (1973) made a critical appraisal of the existence of a cation transport ATPase in roots and concluded that ATPases isolated from the roots of various grasses are not cation specific and are not related to the capability of root cells to absorb ions. However Rungie and Wiskitch (1973) found an ion stimulated ATPase which appeared to be on the surface of the tonoplast and which showed specificity for anions rather than cations. Also Hill and Hill (1973) have reported an ATPase in the salt glands of the halophyte *Limonium* which is stimulated by chloride.

Although evidence for a direct correlation is lacking there has been a lot of work carried out which provides circumstantial evidence for a relationship between cation stimulated ATPases and cation uptake. Fisher, Hanson and Hodges (1970) attempted to correlate rubidium uptake by barley, wheat and oat roots with the activity of ATPase stimulated by rubidium. A ratio of Rb influx/Rb ATPase activity of approximately 0.85 was obtained for all three species. They interpreted this as evidence that sufficient ATPase activity exists to explain the measured uptake rates. However, Baker and Hall (1973) suggest that such an interpretation must be treated with caution as the level of ATPase activity depends on a number of factors such as pH, magnesium concentration and ATP concentration at the membrane. They also point out that the number of ions transported/ATP molecule hydrolysed is not known.

Leonard and Hanson (1972) found that on washing maize roots in either water or salt solutions there was a rise in salt uptake by the tissue which was accompanied by an increase in ATPase activity. Similarly Hill and Hill (1973) found that when *Limonium* plants were taken from a low salt environment and placed in a high concentration of NaCl there was an increase in ATPase activity which was correlated with increased salt secretion.

Perhaps the best evidence for a connection between ATPase activity and salt uptake has come from the work of Leonard and Hodges (1973). They were able to prepare purified extract of the membrane fraction from oat roots which contained more than 75 per cent plasma membrane so they could be fairly certain that they were studying membrane bound ATPase. In the presence of magnesium the ATPase was stimulated five fold by KCl (50 mmol dm^{-3}) but without the addition of magnesium, KCl had little effect. The requirement for magnesium was completely satisfied by manganese but calcium as a replacement for magnesium was ineffective. The inability of calcium to substitute for magnesium was not found for the ATPases from two other membrane fractions, indicating that the plasmalemma ATPase was not contaminated by other membranes. Ouabain had little effect on the ATPase activity and there was no synergistic effect when NaCl and KCl were provided together.

Table 6.1

The effect of various cations on the activity of an ATPase from oat roots. The reaction mixture contained ATP (3 mmol dm^{-3}), tris-MES (33 mmol dm^{-3}), $MgSO_4$ (1.5 mmol dm^{-3}) and monovalent salt 50 (mmol dm^{-3}) pH was 6.0. From Leonard and Hodges (1973).

Cation (given as chloride)	Percentage stimulation
Potassium	100%
Rubidium	87
Sodium	83
Ammonium	80
Caesium	78
Lithium	42
Tris	48
Choline	46
Tetramethylammonium	41

Anions did not have any stimulatory effect but most cations did stimulate the enzyme to varying extents as shown in Table 6.1.

The ATPase did not obey typical Michaelis-Menten kinetics. When KCl stimulation was plotted against the concentration of KCl a discontinuous curve with two phases was obtained (Fig. 6.3). When the data were plotted according to the conventions of Lineweaver-Burk and Eadie-Hofstee, non-linear relationships were observed. Leonard and Hodges suggest that this type of kinetic behaviour may be explained in terms of an enzyme model exhibiting what is termed negative co-operativity. This assumes an enzyme with a number of sub-units in which the binding of the cation to the first sub-unit brings about conformational changes in the other sub-units reducing their affinity for the cation. Thus

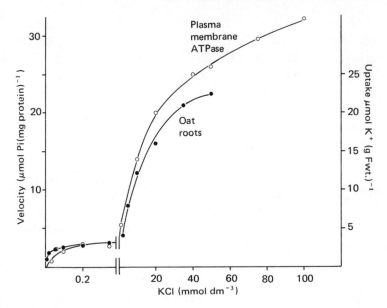

Fig. 6.3

KCl stimulated ATPase activity of the plasma membrane fraction from oat roots plotted together with the rate of ^{42}K uptake from KCl by oat roots. After Leonard and Hodges (1973).

the ability of the enzyme to bind cation becomes progressively reduced.

The curve for ATPase stimulation by KCl shown in Fig. 6.3 is strikingly similar to the biphasic curves obtained for cation uptake in roots by Epstein and others which were discussed in Chapter 5. Also plotted in Fig. 6.3 are some data of Fisher, quoted by Leonard and Hodges, for the uptake of potassium by oat roots. It is clear that the curves for the activity of the plasmalemma ATP and the potassium uptake by the root closely resemble each other and as Leonard and Hodges point out this is another correlation which is consistent with, but not proof of, the implication of ATPases in active salt uptake.

However these results are of considerable importance to our understanding of the biphasic and multiphasic isotherms observed in salt uptake studies. Leonard and Hodges were fairly sure that they were studying one enzyme and so it appears that the uptake systems 1 and 2 for potassium in oat roots are a result of one transport system which has several different affinities as suggested by Nissen (1971) rather than being due to a transport system with two or more different carriers as envisaged by Epstein and his colleagues. Leonard and Hodges did not study the effect of the cations on their ATPase in the concentration range of system 1 but it would be very interesting to know whether the enzyme behaves like the whole root at this level of concentration and if it possesses those characteristics of system 1 such as a high affinity for potassium over sodium and the stimulatory effect of calcium.

It is interesting to note that Leonard and Hodges report only one plasmalemma ATPase which responds to a wide range of monovalent cations and which is unaffected by anions. This lack of specificity to cations suggests that the ATPase is not primarily a carrier of cations across the membrane but that its main role is to make available energy to specific carrier systems by hydrolysis of ATP.

The following scheme is suggested which might explain active transport across the membranes of root cells and which reconciles most of the facts that we have about the process at present. It is shown diagramatically in Fig. 6.4. Cations outside the membrane stimulate ATPase in the membrane to hydrolyse ATP. The energy released is transferred along an electron chain in the membrane which results in the development of a pH gradient across the membrane by the expulsion of H^+ ions. This may also result in the development of a gradient in electrical potential difference with the interior of the membrane negative. It should be pointed out that the gradient of PD is not necessary if a pH gradient is

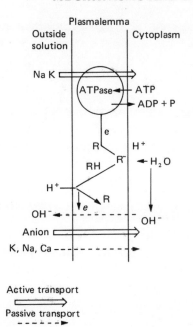

Fig. 6.4

A hypothetical scheme to describe ion
transport across the plasmalemma.

developed but it would explain the evidence for electrogenic
pumps found by some investigators. Anions are taken up
actively by the utilization of the energy in the pH gradient.
How this is brought about is unknown but specific carriers
are probably involved and OH^- is assumed to move out in
exchange for the anion. Active transport of K^+ and Na^+ may
be brought about directly by the ATPase acting as a carrier
but it is also possible that active K^+ and Na^+ transport may be
brought about by carrier systems which are only indirectly
connected to the ATPase and which are similar to those
envisaged for the anions. In this case the membrane ATPase
would have no direct transport role but would act only in the

provision of energy for active transport. A scheme resembling that in Fig. 6.4 has been put forward by Hodges (1973).

Pinocytosis

In animal cells the cell membrane appears to be in a constant state of flux and infolding of the membrane to form vesicles can be observed (Holter, 1960). Evidence from electron micrographs suggests that membrane invagination occurs in plant cells also (Buvat, 1963; Robards and Robb, 1972). Power and Cocking (1970) and Willison, Grout and Cocking (1971) have demonstrated that large molecules such as ferritin and polystyrene latex can be taken into protoplasts by the process of pinocytosis. This is brought about by the engulfing of a volume of the external solution by the infolding of an area of membrane which eventually cuts itself off from the rest of the membrane and forms a vesicle in the cytoplasm (Fig. 6.5).

Sutcliffe (1962) suggested that pinocytosis could be responsible for ion uptake and the transfer of ions across the cytoplasm to the vacuole of plant cells. However, it could be argued that pinocytosis would neither be selective nor be able to produce the large differences in ion concentration between the outside solution and the vacuolar sap which are usually observed. On the other hand it has been suggested that specific binding of ions on to the outside of the membrane could make pinocytosis highly selective and this might also result in an increase of the concentration of the ions in the vesicle (Baker and Hall, 1973).

There is some evidence from the kinetics of uptake by the alga *Nitella* which suggests that pinocytosis may participate in the cytoplasmic phase of ion transport. MacRobbie (1969) found that accumulation of labelled chloride showed a fast and slow component. She suggested that the fast component may be due to the transfer of some of the chloride across the

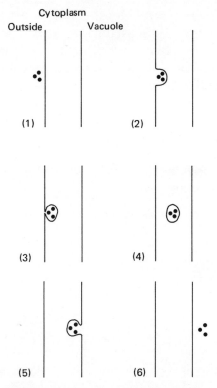

Fig. 6.5

Hypothetical scheme to show how ions
could be transported into the vacuole
by pinocytosis.

cytoplasm in vesicles. Also some workers have reported that
efflux of labelled ions from a number of tissues give washout
curves which do not fit in with the three compartment model
of the cell (Pallaghy, Lüttge and von Willert, 1970; Cram,
1973a and Leigh, Wyn Jones and Williamson, 1973). An
additional cytoplasmic phase is apparent in these results which
might be due to ion transport across the cytoplasm in
vesicles.

Most of the speculations about active transport by pinocytosis lack any substantiating experimental evidence. This is because it is a rather difficult phenomenon to study and most of the evidence for vesiculation is from static electron micrographs. This kind of evidence is essentially qualitative, whereas quantitative information about vesicles is lacking. Data are required about the numbers of vesicles that are formed and the amounts of ions that they can carry because it is virtually certain that *some* ions must find their way into the cell by pinocytosis but what we need to know is whether the process could account for all, or a substantial part of the active ion fluxes observed. Until this kind of information is obtained pinocytosis cannot be considered to be a serious possibility as the mechanism of active salt transport.

7 Transport across the root

Polar movement

An important characteristic of roots is their ability to transport ions and water from the soil to the xylem and hence to the shoot. This centripetal flux presents us with the formidable problem of how ions move, from cell to cell and from tissue to tissue, across the root. There are at least three aspects of this problem that can be singled out for an answer. They are:

(a) What is the mechanism of transport?
(b) What is the pathway taken by the ions?
(c) What determines the polarity of transport?

In practice these three questions overlap because knowledge of the mechanism, for example, will tell us something about the pathway and also will probably go a long way to explain the polarity.

Some data of Ighe and Pettersson (1974) serve to illustrate the ability of the root to transport ions. They followed the uptake of labelled rubidium by wheat roots over a three hour period. Some of their results for relatively well fed plants are

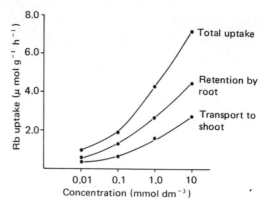

Fig. 7.1

Accumulation and transport of ^{86}Rb by
wheat plants over a three hour period. The
plants were rich in potassium, having been
grown in 0.25 mmol dm^{-3} KCl + 0.25 mmol
dm^{-3} KNO$_3$. Curves plotted from data of
Ighe and Pettersson (1974).

plotted in Fig. 7.1. Whilst a major proportion of the
rubidium taken up by the plants was retained in the root,
approximately 40 per cent was transferred to the shoot. This
fraction remained remarkably constant over the whole range
of external rubidium concentrations. As one might expect
they found that plants previously starved of ions tended to
retain a larger proportion of the uptake in the roots. In
particularly well fed plants however, it is sometimes found
that all the ion taken up is transferred directly to the shoot.

Many plants, on removal of the shoot, exhibit bleeding of
the xylem sap from the cut stump. This phenomenon which
is sometimes called exudation has proved to be a very useful
means of studying polar transport of ions to the xylem. The
flow of xylem sap may continue for several days. The liquid
exuded contains mainly salts with traces of amino acids,
organic acids and growth factors (Kramer, 1969). The total

Table 7.1

Concentrations of the major ions in the external solution and exuding sap of detopped *Ricinus communis* roots together with calculated Nernst potentials and driving forces. Data from Bowling, Macklon and Spanswick 1966.

Ion	External concentration mmol dm^{-3}	Sap concentration mmol dm^{-3}	Nernst potential mV	Driving force kJ mol^{-1}
K$^+$	0.60	5.6	-56	$+ 0.4$
Na$^+$	0.33	0.74	-20	$- 3.2$
Ca^{++}	0.84	10.1	-31	$- 2.1$
Mg^{++}	0.45	3.9	-27	$- 2.5$
NO$_3^-$	1.45	12.0	$+53$	-10.5
Cl$^-$	0.14	1.03	$+50$	-10.2
SO$_4^{--}$	0.69	2.00	$+13$	$- 6.4$
H$_2$PO$_4^-$	0.11	1.38	$+63$	-11.6
HPO$_4^{--}$	0.13	0.56	$+18$	$- 6.9$

mean PD -52 mV

concentration of inorganic ions in the exuded sap is always higher than that in the external solution and some typical data for the xylem sap from excised exuding roots are given in Table 7.1.

The exuding root appears to behave like an osmometer and considerable hydrostatic pressures may be exerted by the exuding sap. The rate of exudation is proportional to the gradient in osmotic potential between the outside solution and the xylem sap (Sabinin, 1925; Arisz, Helder and Van Nie, 1951). This means that the exudation is brought about by the transfer of ions into the xylem and the water follows in response to the osmotic gradient developed. Anderson, Aikman and Meiri (1969) have suggested that the standing gradient hypothesis of Diamond and Bossert (1967) can be used to describe sap flow during exudation. From time to time evidence for a non-osmotic component of water flow has been reported (Van Overbeek, 1942; House and Findlay, 1966) but such claims have never been substantiated.

Exudation therefore demonstrates that the polar transport process is inherent in the root and is largely independent of the influence of the shoot and the transpirational water flux which occurs in the intact plant. It has sometimes been suggested that root exudation may be an artifact caused by decapitation. Certainly the hydrostatic pressures developed during exudation probably bear little resemblance to those in the whole plant. The underlying salt transport however has been found to closely parallel that which occurs in the intact plant (Bowling and Weatherley, 1965).

Root exudation has been studied for over a century and it is still yielding useful information about the process of polar salt transport. The excised exuding root in water culture is a particularly useful system as it enables the composition of the external solution to be changed while at the same time the rate of exudation and the composition of the exuded sap can be followed. Where large volumes of exudate are required whole root systems may be used. A mature root system of tomato for instance will provide 0.5 cm^3 or more sap/hour. A system which provides much less exudate but is easily reproducible where data amenable to statistics are required is that developed by House and his colleagues at the University of East Anglia in the 1960s (House and Findlay, 1966; Anderson and House, 1967). Single roots of *Zea mays* between 8 and 4 cm long are fitted with glass capillaries of uniform bore over their basal ends. The point of insertion of the root is sealed with a lanolin-paraffin mixture. With the root in a suitable medium the rate of exudation is measured by following the rise of the meniscus in the capillary. Salt transport to the exudate is followed by including labelled ions in the medium and counting samples of the sap.

The transport of ions to the xylem exudate bears a close

resemblance to the accumulation process in the cell. There is a relationship between the concentration of an ion in the external solution and its concentration in the exudate (Laine 1934; Bowling and Weatherley, 1964; Hodges and Vaadia, 1964) and the rate of exudation is stimulated by the presence in the medium of certain ions notably chloride and nitrate (Van Andel, 1953). The process is also affected by inhibitors. Van Andel found that cyanide, dinitrophenol, sodium arsenate and other metabolic inhibitors reduced both the uptake of ions from the medium and their flux to the vessels, suggesting at least a partial link between these two stages of transport. Not unexpectedly, low temperatures cause a drastic reduction in exudation rate (Jackson and Weatherley, 1962). Emmert (1966) observed an effect of pH on the process. He found that transport of ^{32}P to the xylem sap in bean was sharply curtailed when the external pH was increased above 6.4.

The accumulation of ions in the xylem sap and the response of exudation to environmental factors strongly suggests that polar transport of ions is an active process. Conclusive evidence for this was obtained by Bowling, Macklon and Spanswick (1966). They applied the Nernst equation to the exudation process is detopped roots of the castor bean (*Ricinus communis*). Using the electrical circuit shown in Fig. 7.2 they measured the electrical potential difference between the exuding sap and the external solution. With the roots in a complete nutrient solution a PD of 50 − 60 mV was found, with the sap negative in relation to the external solution. The Nernst potentials for the major ions were calculated from their concentrations in the medium and the sap. The data are set out in Table 7.1.

The results indicate that all the anions investigated were moving into the xylem against the electrochemical potential gradient, that is, they were actively transported. All the cations, on the other hand, were moving in down the

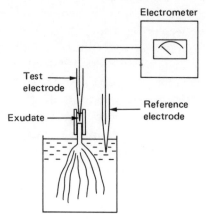

Fig. 7.2

Experimental set-up for measuring the trans-root potential of an excised exuding root system.

electrochemical potential gradient except for potassium which appeared to be close to equilibrium. The polar ion flux therefore appears to be powered by active anion transport with the cations following to maintain the ionic balance.

It can be argued that application of the Nernst equation to the exuding root is a doubtful procedure as there may not be flux equilibrium. There obviously is a net flux to the xylem but it was assumed that the system was close enough to equilibrium for the Nernst equation to be valid. This objection apart, the evidence for the behaviour of the anions can be considered to be conclusive as it is difficult to see how a sustained transport of negative ions against a negative electrical gradient could be brought about by any other process but active transport. However we cannot be so certain about the behaviour of the cations from the data in Table 7.1, as Jennings (1967) and Briggs (1968) have pointed out. It is possible that a large passive movement of a cation could mask a relatively small active component of the transport. There is some evidence for this with potassium and sodium. At low external concentrations when passive transport would be expected to be small, sodium and potassium transport to the xylem in sunflower roots appears to be against the electrochemical gradient (Bowling, 1966; Ansari and Bowling, 1972b).

Since it was first measured, the trans-root potential, as it is often called, has been studied by several investigators using mainly excised corn roots (Davis and Higinbotham, 1969; Shone, 1968, 1969; Dunlop and Bowling, 1971a-c). It behaves very much like the trans-membrane potential of the individual cell. There is a large response to the external concentration of potassium but other ions have relatively little effect. It is rapidly depolarized by cyanide and dinitrophenol and this has led Higinbotham, Graves and Davis (1970) to suggest that electrogenic pumps are involved in polar transport.

How is the trans-root potential related to the trans-membrane potential of the individual cells and where in the root is it located? Answers to these questions should tell us quite a lot about polar transport because the anion pumps are likely to be located where the potential is generated. An attempt to locate the origins of the trans-root potential in maize was made by Dunlop and Bowling (1971c) by comparing the responses of the trans-root and trans-membrane potentials to changes in the external potassium concentration. It can be seen from Fig. 7.3 that both the magnitude of the trans-membrane potential and its response to potassium were greater than for the trans-root potential. This suggested that they were of different origin but later experiments indicated that this was not entirely so.

It was possible to follow both potentials simultaneously on the same root using the experimental set up shown in Fig. 7.4. On changing the external solution from 0.3 to 10 mmol-

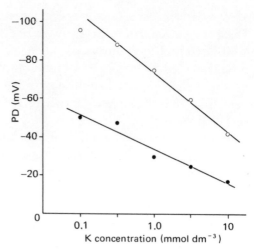

Fig. 7.3

The response of the trans-membrane potential and the trans-root potential of maize roots to external potassium. (○———○); trans-membrane potential, (●———●); trans-root potential. The roots were bathed in KCl + $CaCl_2$ (0.1 mmol dm^{-3}). Data from Dunlop and Bowling (1971a,b).

dm^{-3} KCl there was an immediate response and both potentials were depolarized at about the same rate and to the same extent. The depolarization was virtually complete in 8 seconds, by which time the trans-membrane potential of the epidermal cell had reached the value expected from Fig 7.3. In contrast, the trans-root potential was depolarized by considerably more than expected and almost as much as the trans-membrane potential. However, over a longer period of time the trans-root potential began to rise slowly until it reached the value expected from Fig. 7.3 after 1—2h. The time course of depolarization of the trans-root potential is illustrated by the data in Table 7.2. These changes were

128

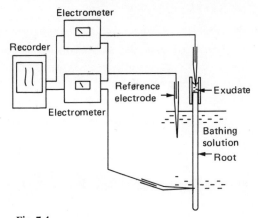

Fig. 7.4

Experimental layout for simultaneously
recording the trans-root and the trans-
membrane potentials in excised maize roots.

interpreted as follows: The initial rapid depolarization was
due to the depolarization of the membrane potentials of the
epidermal and cortical cells but as KCl moved across the root
to the xylem the trans-root potential rose because of the
depolarization of a potential somewhere in the centre of the
root, probably in the stele. The trans-root potential thus
appeared to be made up of the trans-membrane potential of

Table 7.2

Time course of depolarization of the trans-root potential when the
external KCl concentration around a maize root was increased from 0.3
to 10.0 mmol dm^{-3}. Values are means of eight readings ±95 per cent
confidence limits (From Dunlop and Bowling, 1971c).

Time to reach max. depolarization	Maximum depolarization	Final depolarization	Depolarization expected for (a) exudate (b) cell	
24.3 ± 7.8 min	41 ± 7.8 mV	31 ± 9.1 mV	28 mV	45 mV

the outer cells and a much smaller potential in the stele acting in the opposite direction to the outer potential.

The two fold nature of the trans-root potential was confirmed by direct measurements of the PD profile across the root. Thin tertiary roots were mounted intact under the microscope and a micro-electrode moved progressively across the root. It was possible to observe in optical section the tip of the micro-electrode in each of the cells in turn between the epidermis and the xylem. The PD profile obtained in this way for sunflower roots in culture solution is shown in Fig. 7.5. It can be seen that the PD with respect to the

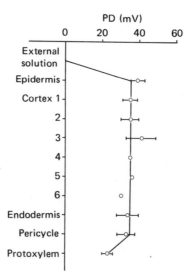

Fig. 7.5

Vacuolar electrical potential differences with the respect to the external solution of the various cells of the intact sunflower root bathed in culture solution. The external potassium concentration was 7.0 mmol dm^{-3}. The bars represent 95% confidence limits. After Bowling (1972).

outside solution was approximately the same for all the living cells of the root at about -35 mV. The PD of the xylem vessels was however lower at -23 mV. This indicates that the trans-root potential is made up of an outer component of -35 mV and an inner component situated at the pericycle-xylem interface of about -12 mV. Dunlop (1973) obtained a similar potential profile for rye grass except that the inner component of the trans-root potential appeared to be at the endodermis-pericycle interface.

It is interesting to consider why there are two components to the trans-root potential. As all the living cells of the root have the same internal potential at a given external potassium concentration there appears to be good electrical continuity between the cells. The electrical resistance of the cell walls is also likely to be very low as they are filled with salt solution, therefore the endodermis must be acting as a resistor to prevent the trans-root PD from being short circuited. In fact it can be argued that the presence of a PD across the root is evidence that the Casparian band is an effective barrier to salt and water movement. The two components of the trans-root PD are due to there being two electrical compartments, an outer one exterior to the endodermis and an inner one inside the endodermis. The outer compartment is bathed by the external solution and the inner one by the xylem sap. The membrane potentials of the cells in the outer and inner compartments will largely depend on the potassium concentrations in the external solution and the xylem sap respectively. Consider the model root shown in Fig.7.6. If we assume it to be a maize root bathed in 1 mmol dm^{-3} KCl then from Fig. 7.3 the trans-membrane potential of the outer cells is -73 mV. Roots of *Zea mays* in 1 mmol dm^{-3} KCl have a concentration of K in the exuding sap of 18.5 m mol dm^{-3} (Dunlop and Bowling, 1971b). The trans-membrane potential of the cells of the stelar compartment should therefore be -35 mV (Fig. 7.3) assuming the membranes to

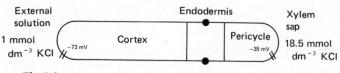

Fig. 7.6

Diagrammatic representation of the root illustrating, the development of the trans-root potential.

behave like those of the outer cells. The trans-root potential is then $-73 - -35 = -38$ mV. The measured value of the trans-root potential for a root in 1 m mol dm^{-3} KCl, given by the lower curve of Fig. 7.3 is -34 mV which is very close to the theoretical value. The trans-root potentials calculated in this way were not found to be significantly different from the measured values for corn roots over a wide range of external concentrations (Dunlop and Bowling, 1971c) Thus providing strong supporting evidence for the two fold nature of the trans-root potential. The implications of this knowledge about the location of the trans-root potential to our understanding of polar transport will be considered later.

The pathway of polar transport-symplasm or free space?

Free space

Salts can penetrate deeply into the root by way of the cell walls. Lavison observed in 1910 that various coloured salts could penetrate the root as far as the endodermis and they appeared to be restricted to the cell walls. Robards and Robb (1972) using more sophisticated techniques have confirmed that the endodermis acts as an effective barrier to cell wall movement. They fed barley roots with uranyl acetate and examined them under the electron microscope. The uranyl ions appeared as electron opaque crystals which were present in the cell walls right up to the Casparian band, but not beyond. However at the root apex where the Casparian band had not yet developed uranyl ions were observed to move into the xylem. Priestley and his collaborators in the 1920s were perhaps the first to understand the significance of the penetration of salts and water in the cell walls as far as the endodermis. They realized that it was not the surface area of the root that was important in salt uptake but the total volume of the cell walls in the cortex (Scott and Priestley, 1928).

Hope and Stevens (1952) were the first to provide some sort of measure of the root volume penetrated by the soil solution. They observed a reversible diffusion of KCl into bean roots. The term coined for this diffusional component was Apparent Free Space (AFS). They calculated that AFS occupied 13 per cent of the tissue. At about the same time Butler (1953) measured the AFS in wheat roots and obtained values of 25 — 34 per cent for Cl and 24 per cent for mannitol, suggesting that diffusional penetration of ions into the root was quite considerable.

There are several different methods for determining AFS and perhaps the simplest is as follows: The root is immersed in a test solution and after a period of say one hour, it is removed and blotted. It is then transferred to a large volume of water and the amount of ion which diffuses out determined. It is assumed that the test solution diffuses into the free space until it reaches the same concentration as that outside. Therefore the amount of diffusable ion recovered divided by its concentration in the test solution gives a measure of the volume occupied by the ion in the root. The proportion of root volume occupied by the free space can then be calculated knowing the fresh weight of the root and the root density. This method only measures the water free space component of the apparent free space. The Donnan free space component has to be determined by cation exchange (see Chapter 2). Some values for AFS obtained by a number of workers are given in Table 7.3.

133

Table 7.3

Estimates of the volume of apparent free space in the root.

Apparent free space %	Tissue	Reference
13	Broad bean	Hope and Stevens, 1952
24.5—33.5	Wheat	Butler, 1953
23	Barley	Epstein, 1955
27.5	Wheat	Kylin and Hylmö, 1957
16—21	Sunflower	Pettersson, 1960
11—14	Wheat	Inglesten and Hylmö, 1961

Hope and Stevens, and Butler considered that the free space volumes they obtained were too large to occupy only the cell wall and were of the opinion that the cytoplasm must also form part of the AFS. This led some workers to the conclusion that the free space extended unimpeded to the xylem and consequently transport to the xylem was a purely passive process (Hylmö, 1953; Kramer, 1957). Levitt (1957) pointed out however, that the values for AFS were very likely inflated because of the carry over of a layer approximately 0.02 mm thick of the test solution on the root surface. Correction for this surface film reduced AFS values from previously published data down to approximately 8 per cent. This corresponded to a value of 11 — 12 per cent for the volume of the cell walls calculated by Butler. Inglesten and Hylmö (1960) removed the surface film by centrifugation and obtained values for AFS of about 11 per cent, thus supporting Levitt's calculations. The experiments of Hope and Dainty (1959) with *Chara australis* described in Chapter 2 appear to confirm the conclusion that the AFS occupies only the cell wall and does not extend into the cytoplasm.

Bernstein and Nieman (1960) used Indian ink to distinguish between superficially retained solution and that entering the free space in bean roots. They found that observed rates of exodiffusion from the free space agreed well with calculated values for cylinders of the same dimension as bean roots, assuming that the free space extended $2/3$ of the way into the root. The physiological data therefore agree very well with the original observations of Lavison.

The symplasm

The evidence that polar transport is an active process means that the pathway taken by the ions on their way to the xylem must involve the crossing of at least one membrane. If the ions traverse the cortex in the cell walls then it will be the membranes of the endodermis that must provide the active step, assuming that the Casparian band is an effective barrier to further cell wall movement. However in the late 1930s when plant physiologists were more concerned with salt accumulation in the vacuole than uptake into the cell wall, another pathway for polar transport was suggested by Crafts and Broyer (1938) and the earlier work of Lavison and Priestley was largely ignored. They postulated that ions were accumulated by the root at the epidermis and were transported inwards to the xylem not by way of the cell walls but via the cytoplasmic continuum which extends from cell to cell through the plasmodesmata. This continuum is called the symplasm or symplast and the theory of Crafts and Broyer is often called the symplasm theory. Therefore we are faced with two possible pathways for polar transport across the cortex, the cell walls and the symplasm. It is generally agreed that the vacuoles of the cortical cells are off the main transport pathway. The main evidence for this comes from some experiments of Broyer (1950). He compared the transport of labelled bromine in tomato roots that were either well fed

135

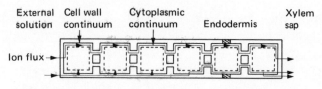

Fig. 7.7

Diagrammatic representation of the root to show the probable pathways taken by ions transported to the xylem.

with bromine or had been deprived of it. In the low bromine plants most of the label was accumulated in the vacuoles and little reached the xylem. In the well fed plants, on the other hand, the labelled bromine was rapidly detected in the xylem sap and appeared to have arrived there without passing through the vacuoles. Some of the possible pathways for ion transport across the root are shown in Fig. 7.7.

The idea that ions can move from cell to cell in the symplasm has attracted considerable interest and support over the years. An early protagonist of the symplasm theory was the Dutch plant physiologist W. H. Arisz. He carried out extensive experiments on the lamina of the aquatic plant *Vallisneria* and showed that polar transport of ions occurs in the symplasm. By analogy a similar transport should occur in the root cortex (Arisz, 1956). Direct evidence for symplasmic transport in the root however has only been obtained comparatively recently. Ginsburg and Ginzburg (1970) have carried out some remarkable experiments on maize roots using what they call 'sleeves'. They separated the cortical cylinder, the 'sleeve' from the stele using a technique first used by Laties and Budd (1964). In the original technique the stele and cortex were separated using a wire stripper but Ginsburg and Ginzburg were able to simply pull them apart. The break between cortex and stele occurred across the outer part of the endodermis so that the sleeve did not include the

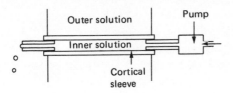

Fig. 7.8

Experimental arrangement used by
Ginsburg.and Ginzburg (1970) to
determine ion fluxes across isolated
cortical sleeves from maize roots.

Casparian band. They measured the fluxes of ^{42}K, ^{22}Na and
^{36}Cl across the cortex using the experimental arrangement
shown in Fig. 7.8. The isotope was placed in either the outer
solution to measure influx across the cortex to the central
cavity or in the inner solution to determine efflux. They
found that 'flux in' equalled 'flux out' for Na^+ and K^+ but for
Cl^- 'flux in' was four times greater than 'flux out'. The PD
across the sleeves was very low (4 or 5 mV) and as the
concentration in the outer and inner solutions were kept the
same they concluded that Cl^- movement inwards was active
and therefore by way of the symplasm.

Further evidence for transport in the symplasm has come
from another direction. It was found that by artificially
increasing the water flux across a detopped sunflower root
the trans-root potential could be increased (i.e. it became
more negative) (Ansari and Bowling, 1972a). This effect
appeared to be linked to an accompanying increase in salt
flux rather than to the water flux itself. The trans-membrane
potential of an epidermal cell and the trans-root potential
were measured simultaneously using an experimental arrangement similar to that shown in Fig. 7.4. On increasing the
water (and salt) flux the trans-root potential altered immediately. The epidermal cell PD also became more negative but
only after a lag of about two minutes. This suggests that on

137

increasing the salt flux the PD between the xylem vessels and the cells surrounding changes immediately and the effect is transmitted outwards from cell to cell back to the epidermis. Assuming that this wave of potential change was due to the change in ion flux the approximate velocity of symplasmic movement could be calculated. It turned out to be 3 cm h^{-1} which is about the same as the rate of protoplasmic streaming in many cells.

It is interesting to note that Epstein and Norlyn (1973) have made some measurements of radial velocities in corn roots. They gave pulses of radioactive Rb and Br at two places along the excised root and obtained two successive peaks of radioactivity in the exuding sap. They were able to calculate the velocities of radial movement of ^{86}Rb and ^{82}Br between the external solution and the xylem and obtained values of 1.8 and 1.4 cm h^{-1} respectively. These velocities are of the same order as that above and compare well with values of $1 - 4$ cm h^{-1} for symplasmic movement of chloride in *Vallisneria* (Arisz and Wiersema, 1966).

Symplasmic movement between cells must be through the plasmodesmata. There is ample evidence for the occurrence of plasmodesmata in most plant tissues including the root cortex. Tyree (1970) made a survey of the information about plasmodesmata and concluded that their frequency ranges from 6×10^9 pores cm^2 in the trumpet cells of *Laminaria digitata* to 1.5×10^8 pores cm^2 in the mature cortical cells of the root of *Allium cepa* (the onion). He calculated that plasmodesmata occupied about 1 per cent of the area in common between adjacent cells in the onion root cortex. The typical radius of the plasmodesmata is 40 nm although it may be as small as 10 nm and as large as 88 nm. The structure of plasmodesmata is a matter of controversy. At present the only point of agreement is that the plasmalemma membrane is continuous from one cell to the next. The membrane forms a cylindrical lining to the pore but the rest of the internal structure is in considerable doubt. The central part of the pore appears to be filled with a central rod about $4 - 6$ nm in diameter but opinion is divided about whether this is part of the endoplasmic reticulum or not. Robards (1971) has interpreted it not as a rod but as a tubule which he calls the desmotubule and which he considers is made up of material originating from the endoplasmic reticulum.

Tyree (1970) has made a theoretical appraisal of symplasmic transport across the root cortex using the thermodynamics of irreversible processes. He concluded that protoplasmic streaming and diffusion could bring about sufficient ion transport in the symplasm to account for the rates of ion flux to the xylem. He envisages bulk movement between cells by diffusion through the plasmodesmata. He calculated that the concentration drop between adjacent cells required for adequate diffusion rates is 0.1 mmol dm^{-3}/cell or about 1 mmol dm^{-3} over the whole cortex. Therefore on theoretical grounds symplasmic transport is quite a feasible system.

With good evidence for salt transport in both the cell walls and the symplasm the question that needs to be answered is not which pathway the salts take across the cortex but rather which of the two pathways is the predominant one. Tanton and Crowdy (1972) have attempted to locate the pathway taken by water in the cortex using lead chelate solutions and precipitating the lead as the sulphide. They found that water appears to move via the cell walls. The endodermis however was found to be a barrier to free movement of lead into the stele. They calculated the velocity of water flow into the xylem through the outer wall of the endodermis as 55.5 nm s^{-1}. Thus if symplasmic transport is regulated by protoplasmic streaming then its velocity will be around $3000-11\,000$ nm s^{-1} that is $50-200$ times faster than the water flux. Therefore it is possible that even though the bulk of the water moves across the cortex in the cell walls the greater part of the ion flux could be transported more efficiently by the symplasm. We require more data on the rates of water and ion movement in the cell walls and the symplasm before this point can be resolved. It is discussed further in Chapter 9.

Profiles of ion distribution across the root

It seems probable that polar transport will give rise to gradients in ion concentration across the root between the epidermis and the stele. Although such gradients are likely to arise first in the symplasm, differences in vacuolar ion concentration between the root cells are also likely to develop. A knowledge of the form of these gradients should tell us something about how they are brought about. For example, a fall in concentration on moving from the epidermis to the stele would suggest that diffusion is the main factor limiting symplasmic transport. Conversely, a rise in ion concentration on moving inwards across the root might indicate that an active transport process is the chief limiting factor. The complete absence of an ionic gradient would suggest that cell to cell transport is a very efficient process and that the overriding factor controlling transport across the root is the rate of removal of ions from the living root cells into the xylem.

A number of studies have been made to determine ion distribution across the root and three main methods have been employed. The most common method has been the use of autoradiography. This was employed by Weigl and Lüttge (1962), Lüttge and Weigl (1962) and Biddulph (1967) to determine ^{45}Ca and $^{35}SO_4$ and by Shone, Clarkson, Sanderson and Wood (1973) for ^{45}Ca and ^{24}Na and by Crossett (1967) who used ^{32}P. Läuchli (1967, 1972) has used the electron probe microanalyser to determine the distribution of potassium. A summary of the main quantitative results obtained in these studies is given in Table 7.4.

Table 7.4

Data obtained by various workers for the distribution of ions in the tissues of the root. All the figures refer to a zone approximately 1 cm from the root tip.

Plant material	Ion	Units	Epidermis	Cortex	Endodermis	Pericycle	Xylem parenchyma	Xylem	Reference
Zea mays	$^{35}SO_4$	Relative	6.9	1.5	1.5	3.5	14.5	18.5	Weigl and Lüttge, 1962
Zea mays	^{32}P	grain density	0.4	3.0	4.4	–	–	3.8	Crossett, 1967
Hordeum vulgare	^{45}Ca		0.43	0.28	0.45	0.68	1.1	0.44	Shone et al., 1973
Zea mays	K	Relative Xray intensity	16	6	10	18	39	4	Läuchli, 1972

The results in Table 7.4 generally show an increase in concentration across the cortex with a peak in the xylem parenchyma surrounding the vessels. There is also a tendency for a relatively high accumulation of label in the epidermis. The significance of these data to polar transport is not clear. It is tempting to suggest that the peaks of radioactivity correspond to sites of active accumulation but the opposite argument, that where there is little accumulation of label, there is the highest rate of transport, is equally tenable. Two disadvantages in both these methods are that it is difficult to prevent movement of ions during preparation of the tissue for scanning and that the data obtained are in relative and not absolute units.

A third method that has recently been employed, is the use of ion specific microelectrodes. This method has the advantage that it measures chemical activity which is a meaningful unit in both physiological and biophysical terms. These microelectrodes work exactly like the common laboratory pH electrode in that they produce a potential difference which is proportional to the chemical activity of the ion to which they are specific. When used in cells the membrane potential of the cell has of course to be subtracted from the reading. To measure profiles of ion distribution in the root the microelectrode is inserted into the tissue using a micromanipulator and viewed with the microscope at quite high magnifications in order to be sure of the location of the sensitive tip. Measurement of the profile of potassium concentration in the maize root using a potassium specific microelectrode is illustrated in Plate 5.

No significant trend was found in potassium activity across the maize root measured in this way (Fig. 7.9). All the vacuoles of the living cells between the epidermis and the xylem were found to have the same potassium concentration which was approximately 110 mmol dm^{-3} (Dunlop and Bowling, 1971a). A similar result was found for sunflower roots

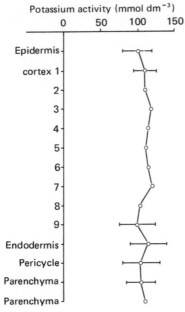

Fig. 7.9

Profile of vacuolar potassium activity across the
root of *Zea mays*. The bathing solution was
1.0 mmol dm^{-3} KCl + 0.1 mmol dm^{-3} CaCl$_2$.
The bars represent 95% confidence limits. After
Dunlop and Bowling (1971a).

(Bowling, 1972) and roots of rye grass (Dunlop 1973). No
peak on potassium activity in the xylem parenchyma was
observed in any of these data and so they contrast with the
results of Läuchli (1972). Läuchli has suggested that this is
because the electron probe analyser detects potassium in
both cytoplasm and vacuole whilst the potassium micro-
electrode measures only the vacuolar activity and there may
be a high accumulation of potassium in the cytoplasm of the
xylem parenchyma.

The results from all the three methods mentioned above
unfortunately show no definite pattern of ion distribution

143

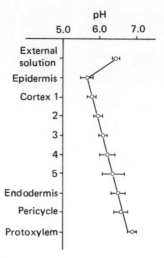

Fig. 7.10

Profile of vacuolar pH across the root of
Helianthus annuus (sunflower). The bars
represent 95% confidence limits. After
Bowling (1973c).

which might throw light on the mechanism of polar
movement from cell to cell. Recently however a micro-
electrode has become available which enables the precise
measurement of intracellular pH. It has been used to measure
the profile of vacuolar pH across the intact root of sunflower
(Bowling, 1973c). Some of the results are shown in Fig. 7.10.
The vacuolar pH was found to rise from less than 6.0 at the
epidermis to nearly 7.0 at the protoxylem. This is quite a
large gradient and it is the first time that a smooth gradient
of ion activity has been observed across the root.

If this pH gradient is reflected in the symplasm then we
might have a basis upon which to explain the polarity of salt
transport as pH is known to have a marked effect on salt
uptake by the root (see Chapter 3). The rate of exudation is
also sensitive to pH as mentioned earlier. pH gradients across

the membranes of mitochondria and chloroplasts play an important role in regulating the direction of salt transport in these organelles (Robertson, 1968) and the possibility of a similar role for H^+ in the bounding membrane of the whole cell cannot be ruled out as already outlined in Chapter 6. If Mitchell's chemi-osmotic hypothesis is applicable to the plasmalemma of root cells then for the sunflower roots of Fig. 7.10, membrane bound ATPase activity in the outer cortical cells is likely to be quite different from that in the stelar cells. Even a small change in the pH gradient may have a profound effect on the rate of active ion transport across the membrane. It is therefore not too difficult to see how the pH gradient in Fig. 7.10 could give rise to unequal rates of salt uptake by the various root cells, resulting in a centripetal ion flux to the xylem. How the pH gradient across the root arises in the first place is quite another question.

Secretion or leakage of salt into the vessels?

Leakage

Crafts and Broyer (1938) in their symplasm theory saw the centripetal transport of ions occurring in three stages:

(1) Active absorption of ions at the epidermis.
(2) Transport through the symplasm down a concentration gradient to the stele.
(3) Leakage of ions from the stelar cells into the dead xylem vessels.

Leakage into the xylem was thought to be a passive process due to the inability of the living stelar cells to retain ions as efficiently as those at the outer surface of the root. They postulated that this would be brought about if there was a deficiency of oxygen in the stele. It was assumed that back leakage of ions from the stele to the outside solution would be prevented by the Casparian band at the endodermis.

145

Table 7.5

Chloride absorption by maize root tissues over a 3h-period from 1.4 mmol dm^{-3} KCl with ^{36}Cl as label. Data of Laties and Budd, 1964.

	uptake cpm/100 mg fresh weight	
Tissue	Fresh	Aged
Stele	484	9600
Cortex	3.840	9620
Intact root	3550	5770

The evidence for points (1) and (2) has already been discussed. Evidence for point (3), leakage of ions into the vessels, was furnished by Laties and Budd (1964). They studied the behaviour of the cortex and stele of corn roots isolated by the wire stripper technique mentioned earlier. An electrical wire stripper was drawn along the root from the base to the tip. It was found that in about half the roots the stele separated clearly and easily from the surrounding tissue.

Their experiments showed that uptake of labelled chloride from dilute KCl solution by freshly isolated steles was much less than that of the isolated cortex (Table 7.5). Aging the tissues in 0.1 mmol dm^{-3} CaSO^4 for 24h resulted in much greater uptake, with the steles showing by far the biggest increase. Conversely fresh steles were found to lose ^{36}Cl at a greater rate than the fresh cortex whilst aged steles were no more leaky than the fresh cortex (Fig. 7.11). They concluded that in the intact root the cortex has a greater ability to take up and retain ions than the intact stele. On isolation, the stele acquires this accumulatory capacity by the activation of a transport system during the aging process.

There has been some controversy over the behaviour of isolated steles and cortices. Yu and Kramer (1967) found that fresh isolated steles from corn roots accumulated ^{32}P as

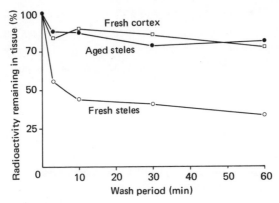

Fig. 7.11

The loss of ^{36}Cl from excised root tissues of *Zea mays* with time. Plotted from data of Laties and Budd (1964).

rapidly as the cortex. Also respiration of isolated steles appeared to be greater than that of the isolated cortex, indicating that the stele is not as metabolically inactive as Crafts and Broyer assumed. Like Laties and Budd, Hall, Sexton and Baker (1971) found that fresh isolated cortices showed higher salt uptake and respiration than fresh steles. However, they attributed the rise in metabolic activity of aged steles to microbial contamination. The large increase in ion absorbtion by the stele on washing was not observed when the tissue was isolated from roots grown under sterile conditions. Hall *et al.* and Laties and Budd used ^{36}Cl, and differences in the behaviour of Cl and P may account for some of the divergent results. The ^{32}P employed by Yu and Kramer may have been metabolically incorporated in the tissues. ^{36}Cl is a more labile ion and so will be more easily lost by the cells.

In a second study, Yu and Kramer found that the steles of intact corn roots can accumulate as much or more ^{86}Rb, ^{36}Cl and ^{32}P than the cortex (Yu and Kramer, 1969).

147

Läuchli, Spurr and Wittkop (1970) and Dunlop and Bowling (1971a) obtained similar results for potassium using quite different techniques and so it appears that the stelar cells do have a similar ability to accumulate ions as the cortex. This is perhaps not surprising as without this ability they would be unable to maintain their vacuolar osmotic potential and would become dehydrated.

Therefore if the stelar cells 'leak' ions to the xylem it must be a controlled leak. Perhaps leak is too strong a word to describe the process. It conjures up the thought of wholesale loss of the ionic contents of the stelar cells. It is more likely that the active uptake/passive loss ratio for the stelar cells is slightly lower than for the cortical cells, thus tilting the balance and resulting in a net inflow of ions to the xylem. This situation is illustrated in Fig. 7.12. The fluxes tending to move ions inward are $2x + \frac{1}{2}x$ and those in the outward direction are $x + x$ so there is a net influx to the xylem of $\frac{1}{2}x$. Notice that the driving force for the inward movement is at the epidermis and that all the cells would show a net accumulation of ions.

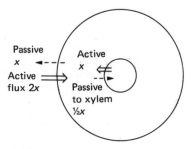

Fig. 7.12

A simple model to explain centripetal ion transport to the xylem. The fluxes at the outer surface of the root are assumed to be greater than those at the inner surface because the surface area of the outer cells is greater.

Further evidence that ion movement into the xylem is a passive step has come from another quarter. We know the profile of electrical potential across the intact root (Fig. 7.5). Knowing the ion concentration in the external solution, the root tissues and the xylem sap, the driving forces on the ions moving into the root from the external solution and from the living root cells to the xylem vessels respectively can be calculated using the Nernst equation. The concentration data for several ions are given in Table 7.6 and the profiles of electrochemical potential for sunflower roots are shown diagrammatically in Fig. 7.13.

It can be seen that K^+, NO_3^- and SO_4^{--} are actively accumulated from the external solution by the root tissues, Ca^{++} and Mg^{++} appear to be moving into the root down the electrochemical potential gradient and are therefore probably not actively accumulated. All the data show without exception that both cations and anions move into the vessels from the surrounding cells down the electrochemical poten-

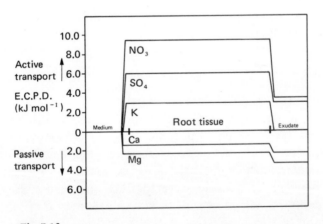

Fig. 7.13

Profiles of electrochemical potential difference (ECPD) for various ions in sunflower roots in culture solution. After Bowling (1973a).

149

Table 7.6

Concentrations of the major ions in the external solution and exuding sap (mmol dm^{-3}) of ion contents (μmol g Fwt.$^{-1}$) of the root tissue between 1 and 2 cm from the root tip of sunflower plants grown in culture solution. Data from Bowling (1973a).

Ion	External Solution	Root	Exudate
K^+	7.0	111	19.0
Ca^{++}	4.0	13	4.0
Mg^{++}	2.0	2.4	2.2
NO_3^-	14.0	150	14.6
SO_4^{--}	2.0	5.8	1.9

tial gradient. Therefore there is no necessity to postulate active transport or secretion into the xylem at this point. Thus the evidence from sunflower roots puts the active transport step at the outer surface of the root and so strongly supports the pump-leak model of Crafts and Broyer. There is even some evidence from sunflower roots for their suggestion of an oxygen gradient across the root (Bowling, 1973b).

Secretion

There have been suggestions from time to time that the final step into the xylem is a secretion (Arisz, 1956; Yu and Kramer, 1969; Läuchli, Spurr and Epstein, 1971). On present knowledge this is difficult to conceive. Plant cells actively accumulate most of the major ions and sodium is the only ion which might be actively secreted. How then can stelar cells which are known to be able to accumulate ions as efficiently as those of the cortex also secrete them? Evidence for active secretion into the xylem is weak. The uncoupler carbonyl-cyanide m-chlorophenylhydrazone (CCCP) has been reported to inhibit the transfer of ions to the xylem in maize roots

(Läuchli and Epstein, 1971) and barley roots (Pitman, 1972). However Baker (1973) has evidence which suggests that the site of action of CCCP is at the cortex and not the stele.

A factor which has an important bearing on whether or not the last step the ions take into the xylem is by secretion is whether the xylem vessels are living or dead. Traditionally plant physiologists have assumed that the vessels in the absorbing zones of the root are dead and empty of living contents. However, there is a body of evidence which suggests that the vessels may be living. Scott (1949) found that in *Ricinus communis* all the vessels were alive up to 8 cm from the root tip and Anderson and House (1967) observed cytoplasm in the vessels of maize up to 6 cm from the tip. This latter finding was confirmed by Higinbotham, Davis, Mertz and Shumway 1973).

Hylmö, in an attempt to explain the origin of root pressure, suggested that exudation from excised roots originates from the immature living vessels at the distal end of the root (Hylmö, 1953). This has become known as the test-tube hypothesis. He suggested that vacuolar sap is liberated upwards from these immature living vessels as the vessels above them mature and their cross walls become perforation plates (Fig. 7.14). The test-tube model implies that exudation and polar transport are identified with the behaviour of the individual cell. This means that the final step in salt transport to the xylem will be a secretory process identical with that responsible for active accumulation in the vacuole. Higinbotham *et al.* (1973) suggested that the trans-root potential is in fact the PD of the immature xylem vessels and that root pressure is identical to the vacuolar hydrostatic pressure.

There are some objections which may be levelled at the test-tube hypothesis. The trans-root potential appears to be more complex than would be expected if it were due to a single cell. In fact the evidence already discussed suggests that the trans-root potential is a tissue phenomenon which is not

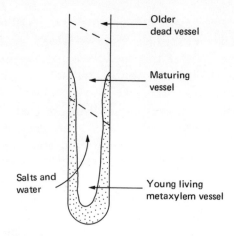

Fig. 7.14

Diagram to illustrate the test-tube model for xylem exudation.

due merely to single cells. Furthermore the evidence for living vessels needs to be extended and confirmed and their role in transfer more clearly defined. Kramer (1969) has questioned whether sap from differentiating xylem vessels is sufficient to account for the volume of flow in exudation. Finally, Tanton and Crowdy (1972) following the distribution of lead in the root, found lead sulphide deposits only in the mature protoxylem elements and not in the cytoplasm of the metaxylem. When the metaxylem had become fully developed lead was found in the lumen of these elements. Therefore in the intact transpiring plant lead movement was confined to mature xylem vessels suggesting that the final step in salt transport across the root is in fact into the dead xylem elements. Whether the same can be said for salt transport in the exuding root remains to be seen.

8 Transport to the aerial parts of the plant

Metabolic relationships between root and shoot

Curtis in 1923 described an experiment in which he ringed small branches of privet just before growth started in the spring. After about a month striking differences between the ringed branches and the neighbouring branches could be seen. The leaves of the ringed branches were small and yellowish-green, while the unringed branches had large green leaves. The leaves on the unringed branches contained 10—100 per cent more nitrogen and ash/unit leaf area than the leaves on the ringed stems. He calculated that 5—20 times more nitrogen moved through the unringed stems than through the ringed stems. This experiment and similar ones carried out at that period suggested that upward transport of inorganic nutrients took place chiefly in the phloem.

In some elegant experiments involving the early use of radioactive isotopes as tracers, Stout and Hoagland (1939) showed that when the xylem and phloem were separated from each other the upward salt transport took place almost entirely in the xylem. Their results have been repeated by many workers since then and it is now generally accepted

153

that the xylem forms the main path for upward movement of salts from the roots to the leaves. How then can we explain Curtis' results?

Ringing plants by removal of the phloem in woody species usually has a great effect on ion uptake by the roots. Koster (1963) ringed soy bean plants and found a rapid fall-off in nitrate and ammonium uptake within a few hours. Castor bean plants showed a decline in potassium uptake almost immediately after ringing (Bowling, 1965). Cooling a length of the stem of sunflower produced a similar fall in salt uptake to that brought about by ringing except that the effect could be reversed in removal of the cooling coil (Bowling, 1968). If plants are left in darkness salt uptake declines but is restored again after a period of illumination as shown in Fig. 8.1. This increase in salt uptake does not occur if the plants are ringed just before the start of the light period. All these results

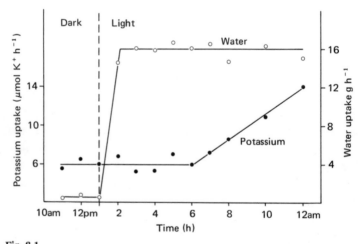

Fig. 8.1

The effect of illumination on the water and potassium uptake of a *Ricinus communis* plant left overnight in the dark. Concentration of potassium in the external solution 0.7 mmol dm^{-3}. Previously unpublished data of the author.

indicate that the uptake of salts is highly dependent on the flow of assimilates to the root in the phloem. This nutritional effect clearly explains Curtis' results and demonstrates that there is a close relationship between the metabolism of the shoot and the root.

Is the decline in salt uptake on ringing due in the first instance to the cessation of carbohydrate supply or is some other vital factor involved? Bange (1965) investigated this point. He found that excised maize roots immediately after excision accumulated potassium at the same rate as attached roots but transport of potassium to the exuding sap was considerably less than potassium transport to the shoot of the intact plant. 1 per cent glucose added to the culture solution considerably enhanced potassium flux into the exuding sap until it was comparable to potassium transfer to the shoot in the intact plant. Hence no special factor other than carbohydrate appears to be required in order to maintain salt transport to the shoot. The uptake itself also appears to be dependent only on sugars. No decline in potassium uptake occurred on ringing sunflower plants if they were provided with sucrose in the outside solution beforehand (Bowling, 1968). Helder found that phosphate uptake by barley plants, reduced in the dark, could be restored by adding glucose (Helder, 1959).

The shoot therefore exerts a major influence on the root through the carbohydrate supply. It may also bring other regulatory factors to bear on the root. Pitman and Cram (1973) have pointed out that the growth rate of the shoot may be an important regulatory influence. They show that there is a linear relationship between potassium transport to the shoot and its relative growth rate. However it is possible to argue that in this case potassium supply to the shoot is regulating the growth rate of the shoot and not vice versa. Also close correlations such as this one do not necessarily imply a causal relationship between the two factors. Conditions which will bring about a rapid shoot growth such as high temperature and light intensity also favour high transpiration rates and high rates of carbohydrate transport which will result in an accelerated rate of salt uptake.

In the situation where carbohydrate supply and the salt supply to the root are adequate it is possible that the shoot may exercise direct control over the transport of salts from the root, otherwise salts moving into the shoot might build up in the intercellular spaces of the leaf. Alternatively any surplus salts in the leaf not taken up by the cells may be retransported back to the root in the phloem. Greenway and Pitman (1965) studied the possibility of retranslocation in barley plants with a high salt status but they found no evidence of large-scale export of potassium from the leaves back to the root. More subtle controls therefore appear to be operating.

We are beginning to find clues to the type of control that might prevail. Some evidence for a regulatory mechanism in plants under water stress has been recently reported by Cram and Pitman (1972). They have found that abscisic acid (ABA) inhibits transport of potassium and chloride from the root to the shoot in maize and barley. They suggest that ABA levels in the leaf may be increased by the osmotic effect of salts building up in the intercellular spaces of the cells of the leaf. Hocking, Hillman and Wilkins (1972) have shown that ABA can be translocated from the leaf to the root. Therefore ABA, by controlling salt transport from the root into the xylem could be the basis of a negative feedback system regulating ion levels in the shoot.

A positive feedback mechanism to regulate nitrate uptake has been proposed by Ben Zioni, Vaadia and Lips (1971). Nitrate is reduced to malate in the leaf, some of which is translocated in the phloem to the root where it is oxidized to $KHCO_3$. $^-HCO_3$ then exchanges with NO_3^- outside (Fig. 8.2). Plants rich in NO_3^- are then able to take in more NO_3^-

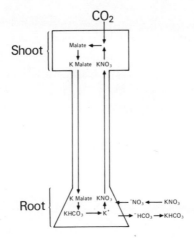

Fig. 8.2

A positive feedback regulatory mechanism
for the control of nitrate uptake. From Ben
Zioni, Vaadia and Lips (1971).

without concomitant uptake of potassium. They have
pointed out that this scheme suggests a useful role for
potassium as a malate and nitrate carrier. Although many
aspects of the scheme have not been experimentally verified
as yet, this kind of approach may prove to be a very valuable
one in aiding our understanding of root-shoot relationships.

The influence of water flux

The question of the existence of a relationship between the
uptake of salts and the uptake of water by the plant has been
discussed since the beginning of plant physiology. De
Saussure in 1804 noticed that plants absorbed proportionally
more water than salts, but his results seem to have been
ignored by the later plant physiologists Sachs and Pfeffer.
They assumed that a close relationship existed between salt
and water absorption, the soil water being drawn *en masse*

157

into the root. Many experiments were carried out in the early part of this century on the problem using plants growing in the field in which transpiration was reduced by shading. The results of this type of experiment have, however, been criticized on the grounds that plants grown in sun and in shade conditions are anatomically and physiologically different and cannot be compared (Steward, 1943; Kramer, 1949). Even in the laboratory some workers have altered transpiration by shading. The experiments of Russell and Shorrocks (1959) and Pettersson (1966) involved reducing transpiration by periods of darkness. The reduced salt uptake they report as being due to a reduction in transpiration is almost certainly due to a fall in the flow of assimilates to the root in the dark. The overriding influence of the assimilate effect over the effect of transpiration is illustrated by the data in Table 8.1. They show that although the plants in the dark transpired at almost twice the rate of those in the light, uptake of rubidium was lower in the dark.

Table 8.1

Uptake of rubidium from 1 mmol dm^{-3} RbCl by two groups of sunflower plants over a 24h period.

One group was placed in the light (5000 lx) under high humidity (100% r.h.). The other group was placed in the dark at low humidity (50% r.h.). From Hatrick and Bowling (1973).

	5000 lx 100% r.h.	Darkness 50% r.h.
Transpiration (g)	5.27	9.53
Dry weight (g)	0.455	0.458
Total Rb$^+$ uptake (mg)	6.110	4.970
Rb$^+$ wt. (mg/gD wt.)	13.429	10.852
$t = 8.44$	$P < 0.01$	

The most acceptable ways to alter transpiration are by altering the relative humidity of the atmosphere surrounding the leaves, whilst the light intensity is kept constant, or, if the plants are grown in water culture by altering the osmotic potential of the solution bathing the roots. Experiments using these two techniques have been carried out by a number of workers and they show beyond any doubt that there is a relationship between transpiration and salt uptake in the intact plant (Broyer and Hoagland, 1943; Hylmö, 1953; Brouwer, 1954; Bowling and Weatherley, 1965; Hooymans, 1969 and others). The results for intact plants have been confirmed with detached roots by artificially inducing a transpirational flux across the root either by applying an external hydrostatic pressure on the surrounding culture solution or by applying suction to the cut stump (Lopushinsky and Kramer, 1961; Jackson and Weatherley, 1962; Jensen, 1962; Pettersson, 1966; Ansari and Bowling, 1972). The picture which emerges from the results of all these experiments is a fascinating but rather complex one. It must be pointed out that there are conditions in which no relationship between water and salt uptake occurs such as when the ion being taken up is present in the external solution at very low concentrations or where its concentration in the root is well below the maximum level (Broyer and Hoagland, 1943; Van den Honert, Hooymans and Volkers, 1955; Bowling, 1968). Under these circumstances most of the ion appears to be side tracked into the vacuoles of the root cells and is unaffected by the transpirational water flux. Under normal conditions when the root is not starved of ions and their supply from the external solution is sufficient, increased transpiration generally results in increased salt uptake as shown by the data of Broyer and Hoagland (1943) reproduced in Table 8.2. The water flux however has little or no effect on the accumulation of ions by the root itself and the salt taken up in response to the

Table 8.2

Potassium uptake by barley plants of different salt status in relation to water uptake. Data of Broyer and Hoagland (1943).

Salt status of plants	Environmental conditions	Water absorbed $cm^3 (g\,Fwt.)^{-1}$	Potassium absorbed $mmol \times 10^2 (g\,Fwt.)^{-1}$
low	low humidity	9.6	10.9
low	high humidity	3.6	10.4
high	low humidity	8.1	5.2
high	high humidity	2.6	3.2

water flux all finds its way to the shoot (Pettersson, 1960; Bowling and Weatherley, 1965). The water flux to the xylem appears to influence the transfer of all the major ions to the shoot although anions like NO_3^-; and PO_4^{---} which are usually rapidly incorporated into organic compounds on entry into the root may be subjected to more overriding biochemical controls (Loughman, 1966; Ben Zioni, Vaadia and Lips, 1971).

There has been considerable controversy about the mechanism of the water dependent ion flux to the shoot. There are two main possibilities, a mass flow of the external solution to the shoot as was tacitly assumed by Sachs and Pfeffer, or an active salt transfer which is somehow speeded up by the water flux. It is of course possible that both these processes could occur simultaneously. The evidence for the first possibility accounting for the whole of the salt transfer to the shoot is virtually negligible. The concentration of the salts in the xylem would always closely reflect that in outside solution if it were drawn in by mass flow, but this is not usually found. Furthermore, exudation is an active process (see Chapter 7) and so it is extremely unlikely that all the salt transferred to the xylem in the intact plant is moved by passive means.

Despite this circumstantial evidence against passive mass flow to the xylem, Hylmö (1953) concluded that $CaCl_2$ transport to the shoot of pea plants was a passive process. He based this conclusion on the finding that there was a relatively large transpirational effect and that the concentration of $CaCl_2$ in the transpiration stream was lower than that in the surrounding solution. He envisaged the mass flow of the external solution to the xylem with the root acting as a sieve by abstracting some of the ions *en route*. In a later paper in which sulphate uptake by wheat was studied it was recognized that there may also be an active component due to exudation (Kylin and Hylmö, 1957).

Quite different results, however, were obtained by the Dutch plant physiologist Brouwer who carried out some elegant experiments on single roots of *Vicia faba* (Brouwer, 1954). He was able to measure water and salt uptake by four distinct zones of the root by enclosing them in small chambers which acted as potometers. A diagrammatic description of his apparatus is shown in Fig. 8.3. He altered the water uptake of the rootlets in the potometers by changing the osmotic potential of the solution around the rest of the root. The effect of increasing the water flux 2–3 times on chloride uptake by the various root zones is shown in Table 8.3. Water and salt uptake data had to be obtained from separate roots but it was clear that water and salt fluxes were not altered in the same proportions when the water uptake was increased. There was a greater increase in water uptake than chloride uptake. Brouwer also studied the effect of 2,4-dinitrophenol on the chloride uptake. His results, given in Table 8.4, show that 10^{-5} mol dm^{-3} DNP virtually cut out the increase in chloride uptake due to the increased water uptake. Water uptake was unaffected by the inhibitor so he could separate the salt and water uptake and thus demonstrate that they are distinct processes.

Hylmö (1955) reinterpreted Brouwer's data maintaining

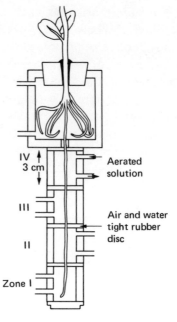

Fig. 8.3

Diagrammatic section of the apparatus used by Brouwer (1954) to measure salt uptake by different root zones.

Table 8.3

Uptake of chloride (μmol) from 5 mmol dm^{-3} CaCl$_2$ over a 24h period by various zones of a rootlet of *Vicia faba*. From Brouwer (1954).

Zone	Low water uptake	High water uptake	Ratio
IV (basal)	12	42	3.5
III	19	52	2.7
II	26	43	1.7
I (apical)	26	33	1.4

Table 8.4

The effect of 2,4-dinitrophenol on chloride uptake
(μmol 24h) by various root zones of a rootlet of
V. faba. From Brouwer (1954).

Zone	Low water uptake	High water uptake	High water uptake $+ 10^{-5}$ mol dm^{-3} DNP
IV	7	22	12
III	11	27	14
II	13	24	9
I	12	29	6

that only part of the water induced salt uptake was inhibited by DNP and that the residual ion uptake due to the enhanced water flux was a passive mass flow component. However, if there is a passive component of salt transport to the xylem it is likely to be very small. Potassium uptake by castor bean plants could be inhibited completely with the antibiotic chloramphenicol whilst water uptake was unaffected (Bowling, 1963). Russell and Shorrocks (1959) found that the ion concentration in the transpiration stream of barley and sunflower may exceed that in the external solution by a factor of over 100. Both these results leave little room for a mass flow component.

The weight of the evidence therefore indicates that in healthy undamaged roots, salt transfer to the xylem, whether water induced or not, is an active process. Thus we are faced with the problem of how the water flow can increase the rate of an active process. Broyer and Hoagland (1943) and Hooymans (1969) have suggested that transpiration might have its effect by reducing the ion concentration in the xylem, thus facilitating salt transfer from the surrounding cells. This is a reasonable possibility where the increase in salt flux does not match the increase in water flux and the concentration of the sap falls progressively. Unfortunately this does not always occur. Jackson and Weatherley (1962) found that increasing the water flux across detopped roots increased the transport of potassium to the xylem without reducing the potassium concentration in the xylem sap. In some experiments the potassium concentration of the sap actually increased on increasing the water flux.

If the concentration of an ion in the xylem sap remains constant at different transpiration rates then a linear relationship between salt transfer to the xylem and transpiration would be expected. This situation was observed for potassium transport in castor bean (Bowling and Weatherley, 1965). Fig. 8.4 shows the effect of increasing the transpir-

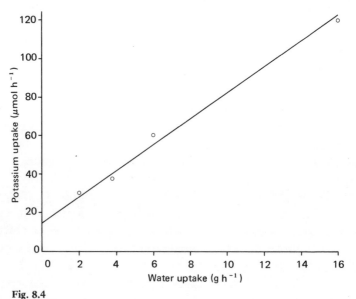

Fig. 8.4

The relationship between water uptake and potassium uptake by a single *Ricinus communis* plant. The external potassium concentration was 0.7 mmol dm^{-3}. From Bowling and Weatherley (1965).

ation rate on potassium uptake by a single plant. Transpiration was increased by lowering the relative humidity around the leaves in steps, in a phytotron. Each step lasted several hours and the points in Fig. 8.4 represent the steady state values for potassium uptake at each transpiration level. The slope of the curve should give the concentration of the xylem sap if all the water-induced potassium uptake is moving to the xylem. This was checked by detopping a number of plants immediately after obtaining curves similar to that in Fig. 8.4 and sampling the exuded sap. A good correlation between the calculated and measured sap concentrations was obtained. The potassium uptake at zero water uptake was accumulated by the root and appeared to be quite independent of the water flux.

The constancy of the xylem sap concentration was also found in detopped sunflower roots in which transpiration was artificially induced by suction (Ansari and Bowling, 1972a). The case of the anions was particularly interesting. As water flow increased, the trans-root potential became more negative and yet the flux of nitrate, for example, increased linearly. The water flux appeared to be accelerating active transport of nitrate to the xylem. The trans-membrane potential of the epidermal cells also became more negative following a lag of about two minutes after the onset of suction. So it appears that active transport across the membranes of the outer cells of the root was increased by the water flux. We have already seen that water movement across the cortex appears to be in the cell walls and so the effect of water flow on active salt transport must be an indirect one. In sunflower roots, transfer of ions into the xylem sap from the surrounding cells is down the electrochemical potential gradient and therefore is likely to be passive (Bowling, 1973a). Therefore it seems highly probable that it is this step which is directly affected by the water although why the sap concentration remains so uniform is difficult to understand.

Increasing the ion flux into the xylem presumably results in an increase in salt transport across the root in the symplasm which in turn brings about increased activity of the ion pumps taking in ions from the outside solution. Inhibitors such as DNP are likely to act on the system at this point cutting off the ion supply to the symplasm but leaving the water flux in the cell walls unaffected. More work is required before we can elucidate the precise nature of this water-salt reaction. We particularly need to understand the crucial step from the symplasm to the xylem.

Transpiration also appears to affect the selectivity of the plant to some ions. It has been found to affect the relative amounts of potassium and sodium transferred to the shoots of barley and mustard (Pitman, 1965, 1966). The roots and shoots of these plants showed a preference for potassium over sodium when the ions were presented together. Increasing the rate of transpiration however reduced the K/Na ratio in the shoot but not in the root (Table 8.5). Pitman considers that this selectivity change probably occurs as a result of the change in water flux past the cation adsorption sites in the cell walls of the root cortex and the xylem.

Table 8.5

The effect of transpiration on potassium and sodium uptake by barley seedlings. The external solution contained 60 mmol dm^{-3} of (KNO_3 + $NaNO_3$) and the K/Na ratio was 1:3. From Pitman (1965).

	Transpiration level	
	low	*high*
Transpiration mg/g F wt. shoot/h	150	300
Shoot potassium uptake μmol/plant	48.8	47.0
Shoot sodium uptake μmol/plant	13.7	19.0
K/Na ratio in shoot	3.6 ± 0.25	2.5 ± 0.1
K/Na ratio in root	3.1 ± 0.1	2.9 ± 0.1

Transport in the xylem

The xylem sap may be obtained for analysis by collecting exudate from the decapitated plants or by extracting the sap from the xylem of woody plants using a vacuum technique (Bollard, 1953). Most of the essential major elements are transported in the xylem as inorganic ions. Nitrogen if present in the external solution as nitrate may be transported along the xylem as NO_3. However, considerable amounts of organic nitrogen compounds may be present in the xylem sap particularly if the plant is fed with ammonium compounds. Table 8.6 shows data obtained by Weissman (1964) for the exudate of sunflower roots fed with different nitrogen sources. In addition to the nitrogen compounds mentioned in Table 8.6 he also detected 11 different amino acids in the exudate.

Iron is not taken up by the root when given as ferric chloride at the near neutral pHs found in the majority of soils, instead it appears to enter the root in a chelated complex. It is readily taken up and transported to the xylem when provided in molecules like ferric ethylenediaminetetra-

Table 8.6

The state of nitrogen in the exudate of sunflower plants fed with different nitrogen sources. From Weissman (1964).

Nitrogen in exudate $\mu g\ N\ cm^{-3}$	External nitrogen source		
	NH_4	$NH_4 + NO_3$	NO_3
Nitrate	—	256	505
Ammonia	169	232	261
Total inorganic N	169	488	766
Total amide	469	393	184
Total amino acid	23	20	19
Total N	661	900	969

acetate (FeEDTA) or diethylenetriaminepenta-acetic acid (DTPA). Wallace and North (1953) used FeEDTA with labelled N and Fe and showed that there was a correspondence between the amounts of the labelled atoms in the original chelate and those in the leaf, indicating that the molecule was absorbed and transported as a whole. There is evidence that calcium may also move up the xylem in chelated form. Jacoby (1966) found that it was taken up and transported to the leaves of intact plants whereas it was retained in the stem of derooted plants. He suggested that the mobility of calcium in the xylem may depend on it forming a complex such as a chelate on its way across the root.

Large amounts of sulphate can be found in the xylem sap indicating that sulphur is transported mainly in inorganic form although methionine has been detected (Pate, 1965). Phosphorus appears to enter the xylem mainly as phosphate (Loughman and Russell, 1957) but traces of organic compounds containing phosphorus have been detected in xylem sap (Morrison 1965; Maizel, Benson and Tolbert, 1956).

Of the minor elements, manganese, copper and zinc were found by Tiffin (1967) to be present in the xylem sap as inorganic ions. These elements do not appear to form complexes in the xylem, as De Kock and Mitchell (1957) found that copper and zinc in particular were not so readily absorbed and transported when chelated.

There appear to be wide seasonal variations in the content of the xylem sap. Bollard (1953) found that in apple trees the ion content of the sap was at its lowest in winter and reached a peak at the time of the opening of the flower buds. Diurnal changes in the sap concentration can also occur with a peak just after noon and a trough around midnight (Vaadia, 1960).

Ions are progressively removed from the xylem sap as they move up to the xylem in the transpiration stream. The sap therefore is more dilute in the stem than it is in the root.

Morrison (1965) found that the concentration of phosphorus in the xylem sap of willow was several times greater in the root than in the stem. The walls of the xylem vessels, in common with other cell walls, are negatively charged and so can retain cations. Bell and Biddulph (1963) observed that tracer calcium was fixed in the stem presumably on the vessel walls and it could be exchanged using $CaCl_2$, $SrCl_2$ or $MgCl_2$. Specific exchange sites for sodium in the root and the base of the stem have been reported by Jacoby (1964, 1965) and Pearson (1967). When a solution containing $1\ \mu mol\ dm^{-3}$ of sodium was pressed through excised stems of *Phaseolus vulgaris* the exudate was found to be almost completely depleted of sodium ions. Jacoby found that sodium retention in these stems was slowed down by metabolic inhibitors indicating that active accumulation by the cells surrounding the vessels as well as exchange on the vessel walls was involved. A similar loss of potassium to the surrounding tissues from the xylem sap of corn roots was found by Meiri and Anderson (1970).

Circulation of ions in the plant

In comparison with ion uptake by roots, ion uptake by leaf cells has been little studied until comparatively recently. Smith and Epstein (1964) found that absorption of rubidium by corn leaf tissue resembles that in excised roots. The process was strongly temperature dependent and they found accumulation ratios of up to 250:1. There is however a major difference between roots and green tissues in that light strongly enhances salt uptake in cells with chloroplasts. Jyung, Wittwer and Bukovac (1965) found that absorption of rubidium in the light by isolated green cells of tobacco exceeded that of the pith cells by almost 100 per cent. Switching on the light has an almost immediate effect on the membrane of cells containing chloroplasts. On illuminating the leaf of *Elodea canadensis* Spanswick (1973) observed a rapid change in the trans-membrane potential. It became more negative by approximately 100 mV within a few minutes of the onset of the light period. The potential could be reduced to its level in the dark by inhibitors such as sodium azide and sodium cyanide and he concluded that there is a light driven electrogenic ion pump in the membrane. The fascinating question raised by the evidence for light stimulated salt uptake is, what is the link between the reactions in the chloroplast and the subsequent effects in the cell membrane? This coupling between light energy and ion uptake in green cells is now a major field of study and the reader is referred to the review by MacRobbie (1970) for further information.

Lateral movement of ions from the xylem vessels into the surrounding tissues including the phloem has been found to occur in stems ever since the experiments of Stout and Hoagland. Most of the major inorganic ions can be detected in the phloem as indicated by the data of Hall and Baker given in Table 8.7. It is not known for certain however which tissues are involved in this lateral transfer between xylem and phloem. Steward and Sutcliffe (1959) have suggested that the cambium may be an important route and in woody tissues the rays are likely to be involved. This is an interesting area which has been little studied and merits further research.

Once in the phloem, ions move downwards in the translocation stream and thus a circulation of ions arises. Mason and Maskell (1931) first suggested the possibility of such a circulation but their experiments were inconclusive and it was left to later workers to demonstrate it. Biddulph and Markle injected ^{32}P through a leaf flap into the cotton plant and found that it moved in the phloem downwards at velocities in excess of 21cm h^{-1} (Biddulph and Markle, 1944). They found that the label could enter the xylem and so move upwards again thus providing the first good evidence

Table 8.7

Inorganic ions in the phloem sap of
Ricinus communis. From Hall and
Baker (1972).

Ion	Concentration mmol dm^{-3}
Potassium	60 −112
Sodium	2−12
Calcium	0.5 − 2.3
Magnesium	4.5 − 5.0
Chloride	10−19
Nitrate	absent
Sulphate	0.25 − 0.5
Phosphate	3.7 − 5.7
Bicarbonate	1.7
Ammonium	1.6
pH	8.0−8.2

for circulation. Helder (1959) confirmed Biddulph and Markle's results and found that 20 − 30 per cent of phosphorus in the leaves of barley could be transferred back to the root at the same time as transfer from the root to the shoot was going on. He also found a similar circulation of rubidium.

Bukovac and Wittwer (1957) applied a number of labelled ions to the leaf and found that they differed in their mobility in the phloem. Rubidium, sodium and potassium were most readily absorbed and proved to be the most mobile within the plant. Divalent cations such as calcium and barium were apparently immobile and there was a large number of ions which possessed an intermediate mobility as shown in Table 8.8.

Biddulph and his colleagues confirmed the immobility of calcium in the plant. They found that calcium taken up by the leaf from the transpiration stream was fixed and younger

Table 8.8

Mobility of ions in the phloem.
From Bukovac and Wittwer (1957).

Mobile	Partially mobile	Immobile
Rb	Zn	Ca
Na	Cu	Sr
K	Mn	Ba
P	Fe	
Cl	Mo	
S		

leaves received no calcium from the older leaves, but had to rely on obtaining it from the transpiration stream. This was in marked contrast to the behaviour of phosphorus. They found that as younger leaves expanded they received ^{32}P from older leaves which therefore lost their radioactivity. Thus radioactivity moved upwards as the plant grew, always being found in the young expanding leaves (Biddulph, Biddulph, Cory and Koontz, 1958), (Biddulph, Cory and Biddulph, 1959).

At first it was thought that calcium did not circulate because it could not be translocated in the phloem. However, recent evidence suggests that it is immobile in the plant for other reasons. Millikan and Hanger (1965) found that foliar applied ^{45}Ca if given alone showed little mobility but if labelled calcium was given with a large dose of stable isotope or with EDTA it became mobile and was found to move all the way down to the root. They concluded that the immobility of calcium is due to the high affinity of leaf tissues for the ion and if it is present in high enough amounts then the surplus will be exported in the phloem.

Another form of circulation is that resulting from the leaching of ions out of leaves by rain or dew. Salts are returned to the soil in this way and may be reabsorbed by the

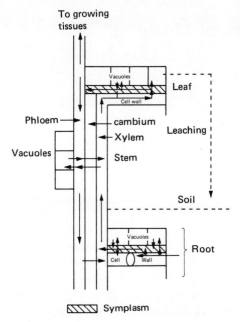

Fig. 8.5

Diagram to show the channels of uptake
and circulation of ions in the plant.

roots. Tukey and Mecklenburg (1964) have shown that
80—90 per cent of the potassium content and 50—60 per
cent of the calcium content can be leached from squash
leaves in 24 h.

Thus most inorganic ions in the plant are in a highly
dynamic state with transport between tissues and between
organs going on continuously. An overall view of the
movements and exchanges that are likely to occur is shown in
Fig. 8.5. The integrated pattern of salt transport illustrated
there emphasizes the fact that salt uptake by the root as a
process cannot be studied satisfactorily without reference to
other processes going on, not only in the root itself, but also
in other parts of the plant.

9 Some conclusions and a look into the future

The individual steps in the catena through which the ions move between the soil and the aerial parts of the plant have been considered in some detail and it is now possible to take an overall view. If we look at the individual processes in the context of the whole plant then hopefully, something new might be revealed about salt transport just as a completed jig-saw puzzle reveals more than its separate parts. Certainly this kind of synthesis should emphasize the gaps in our knowledge even if it reveals little that is new.

Let us consider a young sunflower plant growing in John Innes potting compost at 23°C under a relative humidity of 50 per cent and a steady illumination of 5000 Lux. If we assume its fresh weight to be 3 g then the fresh weight of the shoot will be approximately 2 g and that of the root 1 g. The concentration of potassium in the soil water will be 0.75 mmol dm^{-3} (Ansari and Bowling, 1972b) and under these environmental conditions the rate of potassium uptake will be approximately 25 μmol/gFwt. in 24 h whilst the uptake of water will be 2.0 g/gFwt./24 h (Hatrick and Bowling, 1973). Therefore during a 24 h period our plant will take up 75 μmol of potassium and 6 cm^3 of water.

174

We can apply Passioura's equation to calculate the potassium concentration at the root surface at this level of water flux. If we assume the diffusion coefficient for potassium in the soil is 1.4×10^{-6} cm^2 s^{-1} (Fried and Broeshart, 1967) then putting this value into the equation we find that the potassium concentration at the root surface is 0.69 m mol dm^{-3}. This reveals that 70.5 μmol of potassium arrive at the root surface by diffusion whilst only 4.5 μmol move to the root by mass flow of the soil solution during a 24 h period.

On arriving at the root surface some of the potassium ions may be fixed on the negative sites in the Donnan free space which in our root will have a capacity of approximately 10 μmol (Briggs, Hope and Pitman, 1958). However, other ions particularly calcium are likely to be occupying most of the sites. The majority of the potassium ions will move into the water free space in the cell walls of the cortex which in sunflower was found to occupy about 15 per cent of the root (Pettersson, 1960). In our root this will be 0.15 cm^3.

If we accept the conclusion of Tanton and Crowdy (1972) that water movement across the root cortex is mainly in the cell walls then with a total water flux of 6.0 g the water in the free space will be replaced 40 times in 24 h. Assuming the distance travelled by the water across the cortex to be 1 mm and that the path of the water involves only the radial walls of the cells then the water will move through 100 mm^3. This gives a velocity of 2.5 mm h^{-1}. Obviously this figure does not take full account of the tortuosity of the water pathway across the cortex, a factor which is difficult to assess, but it should be accurate to within an order of magnitude.

The mass flow component of the potassium transport in the free space is $0.69 \times 6 = 4.14$ μmol in 24 h. This is only about 5 per cent of the total uptake of 75 μmol. The diffusion component of transport in the cell walls is difficult to calculate because uptake by the cortical cells will reduce the potassium concentration in the walls to an unknown value. However it is easy to calculate from the diffusion coefficient for potassium that the velocity of diffusion will be between 1 and 2 per cent of the mass flow velocity so that compared with mass flow the diffusion component of free space transport is likely to be negligible.

The potassium concentration in the xylem sap of plants grown under the above conditions was found to be 8.9 mmol dm^{-3} (Ansari and Bowling, 1972b) and this remains fairly constant over a wide range of water fluxes. Therefore the total potassium transport into the xylem is 53.4 μmol in 24 h which is 71 per cent of the potassium uptake. However, this does not take into account potassium recirculating back from the leaves in the phloem which may be transported into the xylem of the root. Our plant will be translocating approximately 11 mg of assimilates to the root in 24 h (Hatrick and Bowling, 1973). Assuming all this dry matter is sucrose and that the concentration of sucrose in the phloem sap is 10 per cent then 0.11 cm^3 of phloem sap will arrive at the root in 24 h. Taking the potassium concentration in the phloem as 100 mmol dm^{-3} (Hall and Baker, 1972) then 11 μmol of potassium will arrive at the root in 24 h. Therefore the total potassium uptake by the root will be 86 μmol and as transport to the xylem is 53.4 μmol then 32.6 μmol will be accumulated by the cytoplasm and vacuoles in 24 h.

It is interesting to note that potassium appears to be moving across the root from the soil water into the xylem at a velocity which is 8.9/0.69 = 12.8 times faster than the water. However this calculation does not take into account the recirculated potassium. Subtracting this would make the concentration in the xylem sap 7.0 mmol dm^{-3} which is still over 10 times more concentrated than the soil water at the root surface. The velocity of water transport is 2.5 mm h^{-1} therefore the velocity of potassium transport into the xylem is 2.5–3.2 cm h^{-1}. This figure agrees very well with the velocities of transport in the symplasm measured by Ansari and Bowling (1972) and Epstein and Norlyn (1973). From this synthesis therefore it seems reasonable to conclude that transport in the symplasm plays the dominant role in salt movement across the root and accounts for 90—95 per cent of the potassium transport across the cortex.

In the root system under consideration it appears that 96 per cent of the potassium arrives at the root surface by diffusion and on being taken into the root is then largely transported in the symplasm probably by cytoplasmic streaming until it reaches the xylem. Transport across the plasmodesmata has not been considered in this synthesis but it does not appear to be a limiting factor as the overall velocity of symplasmic transport is very similar to the velocity of cytoplasmic streaming. This must mean that either streaming, or diffusion across steep concentration gradients is involved (Tyree, 1970). Transport in the xylem is obviously by mass flow and in our plant 20 per cent of the potassium ascending in the stem finds its way eventually into the phloem and is returned to the root. The potassium fluxes revealed by this synthesis are shown diagrammatically in Fig. 9.1.

The major missing link in the catena is the nature of the active transport process across the outer membrane. In our sunflower plant the potassium concentration in the vacuoles will be approximately 60 mmol dm^{-3} and the trans-membrane electrical potential -72 mv (Bowling and Ansari, 1971). This means that potassium is being accumulated against a gradient equivalent to 4.0 kJ mol^{-1} and that the root is expending 0.3 J in 24 h to accumulate potassium. This is not very much in the total energy expenditure of 785 J in that time (Hatrick and Bowling, 1973). How can we study a process which although vital to the plant comprises such a small part of its metabolic activity?

The isolation of the plasmalemma by Hodges and his

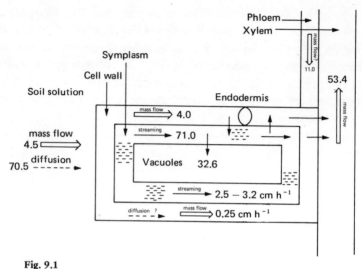

Fig. 9.1

The flow of potassium in a sunflower root (fresh weight 1 g) in potting compost. A synthesis based on experimental data. The fluxes are μmol K^+ 24 h^{-1}.

colleagues is an important step forward and the study of such membrane extracts should yield further information about the transport mechanisms in the membrane. However, we are faced with a dilemma because isolated membranes although containing the transport apparatus are unlikely to transport ions as on removal from the cell their 'sidedness' or polarity will have been destroyed. Therefore studies on the isolated plasmalemma need to be complemented by study of the membrane *in situ*.

A promising approach to the study of the membrane in the intact root has been revealed by the work of Pettersson and his colleagues in Sweden. Pettersson (1966) found an exchangeable fraction of sulphate in sunflower roots that appeared to be related to the active uptake of the ions. Persson (1969) followed up this work and found that this exchangeable fraction was exchangeable with selenate and its magnitude was altered by the external pH. He found a close relationship between the labile bound sulphate and the rate of active uptake of the ion, strongly suggesting that binding to positive sites in the membrane constitutes the first step in active transport. A similar labile bound component of phosphate in the free space of sunflower roots has also been detected (Pettersson, 1970) and in wheat roots there is evidence for a binding of rubidium which also appears to be linked with metabolism (Ighe and Pettersson, 1974).

Transport systems are quite specific and so proteins should constitute the recognition sites as they are the only molecules which have the property to distinguish between different substrates. The proteins involved in the initial binding of an ion prior to active uptake are likely to be located on the outer surface of the membrane and so it should be possible to release them for closer study. In bacteria, proteins that recognize sulphate, galactosides and leucins are released by osmotic shock, a relatively gentle procedure in which bacterial cells are transferred from a sucrose solution of high osmotic potential to a dilute salt solution (Pardee, 1966). About 5 per cent of the total protein of the cell can be released by this treatment. Of course some of the proteins released do not appear to be directly concerned in transport and some binding proteins cannot be removed by osmotic shock. In the red blood cell for example the membrane bound Na-K ATPase is not released by shock treatment.

It would appear that this technique could be very valuable when used on plant cells. However very little use of it has been made by plant physiologists so far. An isolated example of its application to plants is the work of Amar and Reinhold (1973). They found that transferring leaf strips of *Phaseolus vulgaris* from 0.5 mol dm^{-3} sucrose to water at 2°C resulted in the release of about 3.5 per cent of the total cell protein. Osmotic shock reduced the ability of the leaves to take up the amino acid analogue α aminoisobutyric acid. This reduction in uptake was found to be due to a decrease in influx rather than an increase in efflux indicating that the shock treatment had not simply damaged the membrane. They suggest that this shock phenomenon in bean leaves is analogous to that in bacteria.

Work on similar lines has been carried out in the author's laboratory on sunflower roots. Preliminary experiments have shown that osmotic shock releases about 3 per cent of the total protein. This appears to be root protein and not protein from contaminating micro-organisms as the roots were grown under sterile conditions. Rubidium uptake was reduced considerably after osmotic shock treatment as shown by the curves in Fig. 9.2. It seems that here we have a system for

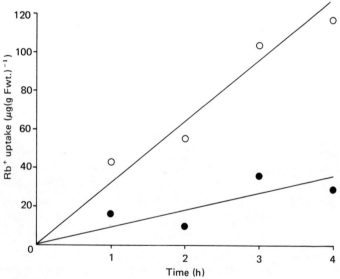

Fig. 9.2

The effect of osmotic shock in the uptake of rubidium by sterile excised roots of *Helianthus annuus*. (○———○); control roots, (●———●); shocked roots. The shock treatment was 0.5 mmol dm^{-3} sucrose followed by water at 2°C. Previously unpublished data of D. J. F. Bowling, A. A. Hatrick and M. Galbraith.

studying active transport proteins in the intact root without the interference of the proteins and general metabolism of the rest of the cell. It would be interesting to know the nature of the proteins released from the root by osmotic shock. Do they for instance include the binding protein for rubidium suggested by the experiments of Ighe and Pettersson and do they have enzymic activity? This approach could prove to be quite a fruitful one in the future.

Another useful approach to the recognition problem is through the study of ionophores. Several antibiotics have been found to stimulate the transport of monovalent cations across a variety of membrane systems including mitochondria (Pressman, 1968). They are collectively called ionophores — ion carrying agents. Pressman has categorized them into two main classes. The first group includes valinomycin, gramicidin and the macrotetralide actins which are neutral and form complexes which acquire the charge of the complexed cation. The second group which includes nigericin contain a carboxyl group which gives the molecule a negative charge.

The ion appears to be liganded to the ionophore by means of induced dipole interaction with oxygen atoms. The ionophore surrounds the ion so that the polar groups are focused towards the interior and the lipophylic groups exposed to the exterior. This makes the ion more mobile in the lipoidal parts of the membrane and valinomycin type antibiotics have been found to facilitate the transport of ions across artificial lipoidal membranes (Liberman and Topaly, 1968). The mechanism of ionophore mediated transport is uncertain but Pressman (1968) considers that the ionophore may help to deliver ions across lipid barriers to the site of the ion pumps.

Hodges, Darding and Weidner (1971) investigated the affect of valinomycin, gramicidin, nonactin and nigericin on potassium uptake by oat roots. They found that valinomycin and nonactin had very little effect. Gramicidin D and

Table 9.1

The effect of various ionophores
($1 \mu g \ cm^{-3}$) on potassium uptake
by oat root segments. From Hodges,
Darding, and Weidner (1971).

Ionophore	Stimulation %
Valinomycin	10
Nonactin	18
Nigericin	99
Gramicidin D	432

nigericin, however, stimulated potassium uptake with gramicidin D being by far the most effective (Table 9.1). The gramicidin stimulated potassium influx was inhibited by metabolic inhibitors including cyanide and DNP. Yoder and Scheffer (1973) have observed that a toxin from the fungus *Helminthosporium carbonum* stimulated the uptake of certain ions by corn roots susceptible to infection by the fungus. Susceptible roots took up more NO_3^-, Na^+ and Cl^- than did control roots. They found that NO_3^- uptake was increased by as much as 163 per cent. The affect of this toxin is specific to susceptible genotypes and it is possible that it is behaving as a highly specific ionophore.

The value of ionophores is that they appear to act in a way which is analogous to the transport proteins in the membrane. If we can isolate proteins from the membrane of the cells of the root which behave like ionophores then we could be well on the way to identifying the elusive carriers. Transport systems cannot have many specific parts because studies with bacteria indicate that few genes are involved in any one of them. Therefore identification of the binding and carrier proteins should hopefully lead us fairly quickly to the identification of the complete transport mechanism within the membrane.

References

Amar, L. and Reinhold, L. (1973) Loss of membrane transport ability in leaf cells and release of protein as a result of osmotic shock. *Plant Physiology,* **51**, 620–625.

Anderson, W. P., Aikman, D. P. and Meiri, A. (1970) Excised root exudation – a standing gradient osmotic flow. *Proceedings of the Royal Society of London,* B **174**, 445–458.

Anderson, W. P. and House, C. R. (1967) A correlation between structure and function in the root of *Zea mays. Journal of Experimental Botany,* **18**, 544–555.

Ansari, A. Q. and Bowling, D. J. F. (1972a) The effect of water and salt fluxes on the trans-root potential of *Helianthus annuus. Journal of Experimental Botany,* **23**, 641–650.

Ansari, A. Q. and Bowling, D. J. F. (1972b) Measurement of the trans-root electrical potential of plants grown in soil. *New Phytologist,* **71**, 111–117.

Arisz, W. H. (1956) Significance of the symplasm theory for transport across the root. *Protoplasma,* **46**, 5–62.

Arisz, W. H., Helder, R. J. and van Nie, R. (1951) Analysis of the exudation process in tomato plants. *Journal of Experimental Botany,* **2**, 257–297.

Arisz, W. H. and Wiersema, E. P. (1966) Symplasmic long distance transport in *Vallisneria* plants investigated by means of autoradiograms. *Proceedings, Koniklijke Nederlandse akademie van Wetenschappen*, c **69**, 223–241.

Arnon, D. I. (1939) The effect of ammonium and nitrate nitrogen on mineral compostion and sap characteristics of barley. *Soil Science* **48**, 295–307.

Arnon, D. I., Fratzke, W. E. and Johnson, C. M. (1942) Hydrogen ion concentration in relation to absorption of inorganic nutrients by higher plants. *Plant Physiology*, **17**, 515–524.

Baker, D. A. (1973) The effect of CCCP on ion fluxes in the stele and cortex of maize roots. *Planta (Berlin)*, **112**, 293–299.

Baker, D. A. and Hall, J. L. (1973) Pinocytosis, ATPase and ion uptake by plant cells. *New Phytologist*, **72**, 1281–1291.

Bange, G. G. J. (1959) Interactions in the potassium and sodium absorption of intact maize seedlings. *Plant and Soil*, **11**, 17–29.

Bange, G. G. J. (1965) Upward transport of potassium in maize seedlings. *Plant and Soil*, **22**, 280–306.

Bange, G. G. J., Tromp, J. and Henkes, S. (1965) Interactions in the absorption of potassium, sodium and ammonium ions in excised barley roots. *Acta Botanica Neerlandica*, **14**, 116–130.

Barber, D. A. (1968) The influence of the microflora on the accumulation of ions by plants. In: *Ecological Aspects of Mineral Nutrition*, ed. Rorison, I. H., Oxford, Blackwell.

Barber, D. A. (1972) 'Dual isotherms' for the absorption of ions by plant tissues. *New Phytologist*, **71**, 255–262.

Barber, D. A. and Frankenburg, U. C. (1971) The contribution of micro-organisms to the apparent absorption of ions by roots grown under non-sterile conditions. *New Phytologist*, **70**, 1027–1034.

Barber, D. A. and Loughman, B. C. (1967) The effect of micro-organisms on the absorption of inorganic nutrients by intact plants. II. Uptake and utilization of phosphate by barley plants grown under sterile and non-sterile conditions. *Journal of Experimental Botany*, **18**, 170–176.

Barber, D. A., Sanderson, J. and Russell, R. S. (1968) Influence of micro-organisms on the distribution in roots of phosphate labelled with phosphorus 32. *Nature (London)*, **217**, 614.

Barber, S. A. (1962) A diffusion and mass flow concept of soil nutrient availability. *Soil Science*, **93**, 39–49.

Barber, S. A., Walker, J. M. and Vasey, E. H. (1962) Principles of ion movement through the soil to the plant root. In: *Transactions of the*

International Society of Soil Science, Commissions II and IV, New Zealand.

Bell, C. W. and Biddulph, D. (1963) Translocation of calcium. Exchange versus mass flow. *Plant Physiology*, 38, 610–614.

Ben Zioni, A., Vaadia, Y. and Lips, S. H. (1971) Nitrate uptake by roots as regulated by nitrate reduction products of the shoot. *Physiologia Plantarum*, 24, 288–290.

Bernstein, L. and Nieman, R. A. Apparent free space of plant roots. *Plant Physiology*, 35, 589–598.

Berry, J. L. and Brock, M. J. (1946) Polar distribution of respiratory rate in the onion root tip. *Plant Physiology*, 21, 542–549.

Bhat, K. K. S. and Nye, P. H. (1973) Diffusion of phosphate to plant roots in soil. *Plant and Soil*, 38, 161–175.

Biddulph, S. F. (1967) A microautoradiographic study of ^{45}Ca and ^{35}S distribution in the intact bean root. *Planta (Berlin)*, 74, 350–367.

Biddulph, O., Biddulph, S. F., Cory, R. and Koontz, H. (1958) Circulation patterns of ^{32}P, ^{35}S and ^{45}Ca in the bean plant. *Plant Physiology*, 33, 293–300.

Biddulph, O., Cory, R. and Biddulph, S. F. (1959) Translocation of calcium in the bean plant. *Plant Physiology*, 34, 512–519.

Biddulph, O. and Markle, J. (1944) Translocation of radiophosphorus in the phloem of the cotton plant. *American Journal of Botany*, 31, 65–70.

Bishop, C., Bayley, S. and Setterfield, G. (1958) Chemical constitution of the primary cell walls of *Avena* coleoptiles. *Plant Physiology*, 33, 283–288.

Black, D. R. and Weeks, D. C. (1972) Ionic relationships of *Enteromorpha intestinalis*. *New Phytologist*, 71, 119–127.

Blevins, D. G., Hiatt, A. J; and Lowe, R. H. (1974) The influence of nitrate and chloride uptake on expressed sap pH, organic acid synthesis, and potassium accumulation in higher plants. *Plant Physiology*, 54, 82–87.

Bollard, E. G. (1953) The use of tracheal sap in the study of apple tree nutrition. *Journal of Experimental Botany*, 4, 363–368.

Bonnett, H. T. (1968) The root epidermis: Fine structure and function. *Journal of Cell Biology*, 37, 199–205.

Bowen, G. D. and Rovira, A. D. (1966) Microbial factor in short term phosphate uptake studies with plant roots. *Nature (London)*, 211, 665–666.

Bowling, D. J. F. (1963) Effect of chloramphenicol on the uptake of salts and water by intact castor oil plants. *Nature (London)*, 200, 284–285.

Bowling D. J. F. (1965) Effect of ringing on potassium uptake by *Ricinus communis* plants. *Nature (London)*, **206**, 317−318.

Bowling, D. J. F. (1966) Active transport of ions across sunflower roots. *Planta (Berlin)*, **69**, 377−382.

Bowling, D. J. F. (1968a) Active and passive ion transport in relation to transpiration in *Helianthus annuus*. *Planta (Berlin)*, **83**, 53−59.

Bowling, D. J. F. (1968b) Translocation of 0°C in *Helianthus annuus*. *Journal of Experimental Botany*, **19**, 381−388.

Bowling, D. J. F. (1972) Measurement of profiles of potassium activity and electrical potential in the intact root. *Planta (Berlin)*, **108**, 147−151.

Bowling, D. J. F. (1973a) The origin of the trans-root potential and the transfer of ions to the xylem of sunflower roots. In: *Ion Transport in Plants*, ed., Anderson, W. P. Academic Press, London.

Bowling, D. J. F. (1973b) Measurement of a gradient of oxygen partial pressure across the intact root. *Planta (Berlin)*, **111**, 323−328.

Bowling, D. J. F. (1973c) A pH gradient across the root. *Journal of Experimental Botany*, **24**, 1041−1045.

Bowling, D. J. F. (1974) Measurement of intracellular pH in roots using a H^+ sensitive microelectrode. In: *Membrane Transport in Plants*, ed. Zimmermann, U. and Dainty, J. Springer Verlag, Berlin.

Bowling, D. J. F. and Ansari, A. Q. (1971) Evidence for a sodium influx pump in sunflower roots. *Planta (Berlin)*, **98**, 323−329.

Bowling, D. J. F., Macklon, A. E. S. and Spanswick, R. M. (1966) Active and passive transport of the major nutrient ions across the root of *Ricinus communis*. *Journal of Experimental Botany*, **17**, 410−416.

Bowling, D. J. F. and Weatherley, P. E. (1964) Potassium uptake and transport in roots of *Ricinus communis*. *Journal of Experimental Botany*, **15**, 413−421.

Bowling, D. J. F. and Weatherley, P. E. (1965) The relationships between transpiration and potassium uptake in *Ricinus communis*. *Journal of Experimental Botany*, **16**, 732−741.

Broeshart, H. (1962) Cation absorption and adsorption by plants. In: *Isotopes in the study of soils and plant nutrition*. International Atomic Energy Association, Vienna.

Briggs, G. E. (1968) Passive movement of water and solutes in to the absorbing region of the root. *Journal of Experimental Botany*, **19**, 486−488.

Briggs, G. E., Hope, A. B. and Pitman, M. G. (1958) Exchangeable ions in beetroot discs at low temperatures. *Journal of Experimental Botany*, **9**, 128−141.

Brown, R. and Broadbent, D. (1950) The development of cells in the growing zones of the root. *Journal of Experimental Botany*, **1**, 249–263.

Brouwer, R. (1954) The regulating influence of transpiration and suction tension on the water and salt uptake by roots of intact *Vicia faba* plants. *Acta Botanica Neerlandica*, **3**, 264–312.

Broyer, T. C. and Hoagland, D. R. (1943) Metabolic activities of roots and their bearing on the relation of upward movement of salts and water in plants. *American Journal of Botany*, **30**, 261–273.

Broyer, T. C. (1950) Further observations on the absorption and translocation of inorganic solutes using radioactive isotopes with plants. *Plant Physiology*, **25**, 367–376.

Bukovac, M. J. and Wittwer, S. H. (1957) Absorption and mobility of foliar applied nutrients. *Plant Physiology*, **32**, 428–435.

Butler, G. W. (1953) Ion uptake by young wheat plants. II. The 'apparent free space' of wheat roots. *Physiologia Plantarum*, **6**, 617–635.

Buvat, R. (1963) Electron microscopy of plant protoplasm. *International Review of Cytology*, **14**,41–155.

Canning, R. E. and Kramer, P. J. (1958) Salt absorption and accumulation in various regions of the root. *American Journal of Botany*, **45**, 378–382.

Clarkson, D. J., Robards, A. W. and Sanderson, R. (1971) The tertiary endodermis in barley roots: Fine structure in relation to radial transport of ions and water. *Planta (Berlin)*, **96**, 292–305.

Crafts, A. S. and Broyer, T. C. (1938) Migration of salts and water into xylem of the roots of higher plants. *American Journal of Botany*, **25**, 529–535.

Cram, W. J. (1968) Compartmentation and exchange of chloride in carrot root tissue. *Biochemica et Biophysica Acta*, **163**, 339–353.

Cram, W. J. (1973a) Chloride transport in vesicles: Implications of colchicine effect on Cl influx in *Chara* and Cl exchange kinetics in maize root tips. In: *Ion Transport in Plants*, ed. Anderson, W. P. Academic Press, London.

Cram, W. J. (1973b) Internal factors regulating nitrate and chloride influx in plant cells. *Journal of Experimental Botany*, **24**, 328–341.

Cram, W. J. and Pitman, M. G. (1972) The action of abscisic acid on ion uptake and water flow in plant roots. *Australian Journal of Biological Sciences*, **25**, 1125–1132.

Crossett, R. N. (1967) Autoradiography of ^{32}P in maize roots. *Nature (London)*, **213**, 312–313.

Crooke, W. M. (1964) The measurement of the cation exchange capacity of plant roots. *Plant and Soil*, **21**, 43–49.

Crooke, W. M. and Knight, A. H. (1962) An evaluation of published data on the mineral composition of plants in the light of the cation exchange capacities of their roots. *Soil Science*, 93, 365—373.

Crooke, W. M. and Knight, A. H. (1971) Root cation exchange capacity and organic acid content of tops as indices of varietal yield. *Journal of the Science of Food and Agriculture*, 22, 389—392.

Curtis, O. F. (1923) The effect of ringing a stem on the upward transfer of nitrogen and ash constituents. *American Journal of Botany*, 10, 361—382.

Curtis, O. F. and Clark, D. G. (1950) *An introduction to plant physiology*. McGraw Hill, New York.

Daft, M. J. and Nicholson, T. H. (1966) The effect of *Endogene* mycorrhiza on plant growth. *New Phytologist*, 65, 343—350.

Dainty, J. (1963) Water relation of plant cells. *Advances in Botanical Research*, 1, 279—326.

Dainty, J. and Hope, A. B. (1959) Ionic relations of cells of *Chara australis*. I. Ion exchange in the cell wall. *Australian Journal of Biological Sciences*, 12, 395—411.

Davis, R. F. and Higinbotham, N. (1969) Effects of external cations and respiratory inhibitors on electrical potential of the xylem exudate of excised corn roots. *Plant Physiology*, 44, 1383—1392.

DeKock, P. C. and Mitchell, R. L. (1957) Uptake of chelated metals by plants. *Soil Science*, 84, 55—62.

De Saussure, N. T. (1804) Recherches chimiques sur la végétation. Madame Huzard, Paris.

Devaux, H. (1916) Action rapide des solution salines sur les plantes vivantes; déplacement réversible d'une partie des substances basiques contenues dans la plante. *Comptes rendus de l'Académie des Sciences de Paris*, 162, 561—563.

Diamond, J. M. and Bossert, W. H. (1967) Standing gradient osmotic flow. A mechanism for coupling of water and solute transport in epithelia. *Journal of General Physiology*, 50, 2061—2083.

Dittmer, H. J. (1937) A quantitative study of the roots and root hairs of a winter rye plant. *American Journal of Botany*, 24, 417—420.

Dittmer, H. J. (1948) A comparitive study of the number and length of roots produced in nineteen angiosperm species. *Botanical Gazette*, 109, 354—358.

Dittmer, H. J. (1949) Root hair variations in plant species. *American Journal of Botany*, 36, 152—155.

Dodds, J. J. A. and Ellis, R. J. (1966) Cation stimulated adenosine triphosphatase activity in plant cell walls. *Biochemical Journal,* **101**, 31–32.

Drake, M., Vengris, J. and Colby, W. G. (1951) Cation exchange capacity of plant roots. *Soil Science,* **72**, 139–147.

Drew, M. C. and Nye, P. H. (1969) The supply of nutrient ions by diffusion to plant roots in soil. *Plant and Soil,* **31**, 407–424.

Drew, M. C., Vaidyanathan, L. V. and Nye, P. H. (1967) Can soil diffusion limit the uptake of potassium by plants? In: *Soil Chemistry and Fertility,* ed. Jacks, G. V. International Society of Soil Science, Aberdeen.

Du Buy, H. C., Woods, M. W. and Lackey, M. D. (1950) Enzymic activities of isolated,normal and mutant mitochondria and plastids of higher plants. *Science,* **111**, 572–574.

Dunlop, J. (1973) The transport of potassium to the xylem exudate of rye grass. 1. Membrane potentials and vacuolar potassium activities in seminal roots. *Journal of Experimental Botany,* **24**, 995–1002.

Dunlop, J. and Bowling, D. J. F. (1971a) The movement of ions to the xylem exudate of maize roots. 1. Profiles of membrane potential and vacuolar potassium activity across the root. *Journal of Experimental Botany,* **22**, 434–444.

Dunlop, J. and Bowling, D. J. F. (1971b) The movement of ions to the xylem exudate of maize roots. 2. A comparison of the electrical potential and electrochemical potentials of ions in the exudate and in the root cells. *Journal of Experimental Botany,* **22**, 445–452.

Dunlop J. and Bowling, D. J. F. (1971c) The movement of ions to the xylem exudate of maize roots. 3. The location of the electrical and electrochemical potential differences between the exudate and the medium. *Journal of Experimental Botany,* **22**, 453–464.

Edwards, D. G. (1968) Cation effects of phosphate absorption from solution by *Trifolium subterraneum. Australian Journal of Biological Sciences,* **21**, 1–11.

Emmert, F. (1966) Centripetal ion penetration in roots as a primary endogenous conditioner of plant top nutrition. *Radioecological Concentration Processes,* an International Symposium. pp. 383–389, Pergamon Press, Oxford.

Epstein, E. (1953) Mechanism of ion absorption by roots. *Nature (London),* **171**, 83–84.

Epstein, E. (1955) Passive permeation and active transport of ions in plant roots. *Plant Physiology,* **30**, 529–535.

Epstein, E. (1961) The essential role of calcium in selective cation transport by plant cells. *Plant Physiology,* **17**, 682–685.

Epstein, E. (1966) Dual pattern of ion absorption by plant cells and by plants. *Nature (London)*, **212**, 1324—1327.

Epstein, E. (1968) Micro-organisms and ion absorption by roots. *Experientia*, **24**, 616—617.

Epstein, E. (1972a) Ion absorption by roots: The role of micro-organisms. *New Phytologist*, **71**, *873—874*.

Epstein, E. (1972b) Mineral nutrition of plants: principles and perspectives. Wiley, New York.

Epstein, E. (1973) Mechanisms of ion transport through plant cell membranes. *International Review of Cytology*, **34**, 123—168.

Epstein, E. and Hagen, C. E. (1952) A kinetic study of the absorption of alkali cations by barley roots. *Plant Physiology*, **27**, 457—474.

Epstein, E. and Norlyn, J. D. (1973) The velocities of ion transport into and through the xylem of roots. *Plant Physiology*, **52**, 346—349.

Epstein, E. and Rains, D. W. (1965) Carrier mediated cation transport in barley roots: Kinetic evidence for a spectrum of active sites. *Proceedings of the National Academy of Sciences (USA)*, **53**, 1320—1324.

Erickson, L. C. (1946) Growth of tomato roots as influenced by oxygen in the nutrient solution. *American Journal of Botany*, **33**, 551—561.

Erickson, R. O. and Sax, K. B. (1956) Elemental growth rate of the primary root of *Zea mays*. *American Philosophical Society Proceedings*, **100**, 487—489.

Esau, K. (1965) *Plant Anatomy*. Wiley, New York.

Eshel, A. and Waisel, Y. (1972) Variations in sodium uptake along primary roots of corn seedlings. *Plant Physiology*, **49**, 585—589.

Eshel, A. and Waisel, Y. (1973) Heterogeneity of ion uptake mechanisms along primary roots of corn seedlings. In: *Ion Transport in Plants*, ed. Anderson, W. P. Academic Press, London.

Etherton, B. (1963) Relationship of cell trans-membrane electropotentials to potassium and sodium accumulation ratios in oat and pea seedlings. *Plant Physiology*, **38**, 581—585.

Etherton, B. (1968) Vacuolar and cytoplasmic potassium concentrations in pea root in relation to cell-to-medium electrical potentials. *Plant Physiology*, **43**, 838—840.

Etherton, B. and Higinbotham, N. (1960) Trans-membrane potential measurements of cells of higher plants as related to salt uptake. *Science*, **131**, 409—410.

Fahn, A. (1967) *Plant Anatomy*. Pergamon, Oxford.

Fisher, J. D., Hansen, D. and Hodges, T. K. (1970) Correlation between ion fluxes and ion-stimulated adenosine triphosphatase activity of plant roots. *Plant Physiology,* **46**, 812—814.

Fisher, J. and Hodges, T. K. (1969) Monovalent ion-stimulated ATPase from oat roots. *Plant Physiology,* **44**, 385—390.

Fried, M. and Broeshart, H. (1967) The soil-plant system. Academic Press, New York.

Gerdemann, J. W. (1964) The effect of mycorrhiza on the growth of maize. *Mycologia,* **56**, 342—349.

Gerson, D. F. and Poole, R. J. (1971) Anion absorption by plants; A unary interpretation of 'dual mechanisms'. *Plant Physiology,* **48**, 509—511.

Gerson, D. F. and Poole, R. J. (1972) Chloride accumulation by mung bean root tips: A low affinity active transport system at the plasmalemma. *Plant Physiology,* **50**, 603—607.

Ginsburg, H. (1972) Analysis of plant root electropotentials. *Journal of Theoretical Biology,* **37**, 389—412.

Ginsburg, H. and Ginzburg, B. Z. (1970) Radial water and solute flows in roots of *Zea mays.* II. Ion fluxes across root-cortex. *Journal of Experimental Botany,* **21**, 593—604.

Goldman, D. E. (1943) Potential, impedance, and rectification in membranes. *Journal of General Physiology,* **27**, 37—60.

Goodwin, R H. and Avers, C. J. (1956) Studies on roots, III. An analysis of root growth in *Phleum pratense* using photomigrographic records. *American Journal of Botany,* **43**, 479—487.

Greenway, H. and Pitman, M. G. (1965) Potassium retranslocation in seedlings of *Hordeum vulgare. Australian Journal of Biological Sciences,* **18**, 235—247.

Gregory, F. G. and Woodford, H. K. (1939) An apparatus for the study of the oxygen, salt and water uptake of various zones of the root with some preliminary results with *Vicia faba. Annals of Botany, New Series,* **3**, 147—154.

Greuner, N. and Neumann, J. (1966) An ion stimulated ATPase from bean roots. *Physiologia Plantarum,* **19**, 678—682.

Gunning, B. E. S. and Pate, J. S. (1969) 'Transfer cells', plant cells with wall ingrowths, specialised in relation to short distance transport of solutes. Their occurrence, structure and development. *Protoplasma,* **68**, 107—133.

Gutknecht, J. and Dainty, J. (1968) Ionic relations of marine algae. *Oceanography and Marine Biology, Annual Review,* **6**, 163—200.

Haberlandt, G. (1914) *Physiological plant anatomy.* Macmillan, London.

Hall, J. L. (1969) Localisation of cell surface adenosine triphosphatase activity in maize roots. *Planta (Berlin)*, **85**, 105–107.

Hall, J. L. (1971) Further properties of adenosine triphosphatase and β glycerophosphatase from barley roots. *Journal of Experimental Botany*, **22**, 800–808.

Hall, J. L., Sexton, R. and Baker, D. A. (1971) Metabolic changes in washed, isolated steles. *Planta (Berlin)*, **96**, 54–61.

Handley, R. and Overstreet, R. (1955) Respiration and salt respiration by excised barley roots. *Plant Physiology*, **30**, 418–426.

Harley, J. L. (1969) *Biology of mycorrhiza*. Leonard Hill, London.

Harley, J. L. and Brierley, J. K. (1954) Uptake of phosphate by excised mycorrhizal roots of the beech. VI. Active transport of phosphorus from the fungal sheath into the host tissue. *New Phytologist*, **53**, 240–252.

Harley, J. L. and McCready, C. C. (1952) The uptake of phosphate by excised mycorrhizal roots of the beech. II. Distribution of phosphorus between the host and the fungus. *New Phytologist*, **51**, 56–64.

Hatch, A. B. (1937) The physical basis of mycotrophy in the genus *Pinus*. *Black Rock Forest Bulletin*, **6**.

Hatrick, A. A. and Bowling, D. J. F. (1973) A study of the relationship between root and shoot metabolism. *Journal of Experimental Botany*, **24**, 607–613.

Heimsch, C. (1951) Development of vascular tissues in barley roots. *American Journal of Botany*, **38**, 523–537.

Helder, R. J. (1959) Exchange and circulation of ions in young intact plants. In: *Radioisotopes in Scientific Research* IV. Pergamon Press, London.

Helder, R. J. and Boerma, J. (1969) An electron microscopical study of the plasmodesmata in the roots of young barley seedlings. *Acta Botanica Neerlandica*, **18**, 99–107.

Hewitt, E. J. (1966) *Sand and water culture methods used in the study of plant nutrition*. Commonwealth Agricultural Bureaux.

Higinbotham, N. (1973) The mineral absorption process in plants. *The Botanical Review*, **39**, 15–69.

Higinbotham, N., Davis, R., Mertz, D. and Shumway, L. (1973) Some evidence that radial transport in maize roots is into living vessels. In: *Ion Transport in Plants*, ed. Anderson, W. P. Academic Press, London.

Higinbotham, N., Etherton, B. and Foster, R. J. (1964) Effect of external K^+, NH_4^+, Na^+, Ca^{++}, Mg^{++} and H^+ ions on the cell trans-membrane potential of *Avena* coleoptile. *Plant Physiology,* **39,** 196–203.

Higinbotham, N., Etherton, B. and Foster, R. J. (1967) Mineral ion contents and cell trans-membrane electro-potentials of pea and oat seedling tissue. *Plant Physiology,* **42,** 37–46.

Higinbotham, N., Graves, J. S. and Davis, R. F. (1970) Evidence for an electrogenic ion transport pump in cells of higher plants. *Journal of membrane Biology,* **3,** 210–222.

Hill, B. S. and Hill, A. E. (1973) Enzymic approaches to chloride transport in the *Limonium* salt gland. In: *Ion Transport in Plants,* ed. Anderson, W. P. Academic Press, London.

Hind, G. and Jagendorf, A. T. (1965) Light scattering changes associated with the production of a possible intermediate in photophosphorylation. *Journal of Biological Chemistry,* **240,** 3195–3201.

Hirata, H. and Mitsui, S. (1965) Role of calcium in potassium uptake by plant roots. *Plant and Cell Physiology,* **6,** 699–709.

Hoagland, D. R. and Broyer, T. C. (1936) General nature of the process of salt accumulation by roots with description of experimental methods. *Plant Physiology,* **11,** 471–507.

Hoagland, D. R. and Broyer, T. C. (1940) Hydrogen-ion effects and the accumulation of salt by barley roots as influenced by metabolism. *American Journal of Botany,* **27,** 173–185.

Hoagland, D. R. and Davis, A. R. (1929) The intake and accumulation of electrolytes by plant cells. *Protoplasma,* **6,** 610–629.

Hocking, T. J., Hillman, J. R. and Wilkins, M. B. (1972) Movement of abscisic acid in *Phaseolus vulgaris* plants. *Nature (London),* **235,** 124–125.

Hodges, T. K. (1973) Ion absorption by plant roots. *Advances in Agronomy,* **25,** 163–207.

Hodges, T. K., Darding, R. L. and Weidner, T. (1971) Gramicidin D stimulated influx of monovalent cations into plant roots. *Planta (Berlin),* **97,** 245–256.

Hodges, T. K. and Vaadia, Y. (1964) The kinetics of chloride accumulation and transport in exuding roots. *Plant Physiology,* **39,** 490–494.

Holter, H. (1960) Pinocytosis. *International Review of Cytology,* **8,** 481–504.

Hooymans, J. J. M. (1964) The role of calcium in the absorption of anions and cations by excised barley roots. *Acta Botanical Neerlandica,* **13,** 507–540.

Hooymans, J. J. M. (1968) Effect of the counter-ion on the uptake of potassium in excised barley roots. *Acta Botanica Neerlandica.* **17**, 313−319.

Hooymans, J. J. M. (1969) The Influence of the transpiration rate on uptake and transport of potassium ions in barley plants. *Planta (Berlin),* **88**, 369−371.

Hope, A. B. and Stevens, P. G. (1952) Electrical potential differences in bean roots and their relation to salt uptake. *Australian Journal of Scientific Research,* **B5**, 335−343.

House, C. R. and Findlay, N. (1966) Water transport in isolated maize roots. *Journal of Experimental Botany,* **17**, 344−354.

Hurd, R. G. and Sutcliffe, J. F. (1957) An effect of pH on the uptake of salt by plant cells. *Nature (London),* **180**, 233−235.

Hylmö, B. (1953) Transpiration and ion absorption. *Physiologia Plantarum,* **6**, 333−405.

Hylmö B. (1955) Passive components in the ion absorption. I. The zonal ion and water absorption in Brouwer's experiments. *Physiologia Plantarum,* **8**, 433−449.

Ighe, U and Pettersson, S. (1974) Metabolism-linked binding of rubidium in the free space of wheat roots and its relation to active uptake. *Physiologia Plantarum,* **30**, 24−29.

Inglesten, B. and Hylmö, B. (1961) Apparent free space and surface film determined by a centrifugation method. *Physiologia Plantarum,* **14**, 157−170.

Jackman, R. H. (1965) The uptake of rubidium by the roots of some graminaceous and leguminous plants. *New Zealand Journal of Agricultural Research,* **8**, 763−777.

Jackson, J. E. and Weatherley, P. E. (1962) The effect of hydrostatic pressure gradients on the movement of potassium across the root cortex. *Journal of Experimental Botany,* **13**, 128−143.

Jacobson, L., Overstreet, R. and Carlson, R. M. (1957) The effect of pH and temeprature on absorption of potassium and bromide by barley roots. *Plant Physiology,* **32**, 658−662.

Jacoby, B. (1964) Function of bean roots and stems in sodium retention. *Plant Physiology,* **39**, 445−449.

Jacoby, B. (1965) Sodium retention in excised bean stems. *Physiologia Plantarum,* **18**, 730−739.

Jacoby, B. (1966) Ascent of calcium in intact and de-rooted bean plants. *Nature (London),* **211**, 212.

Jagendorf, A. T. and Uribe, E. (1966) ATP formation caused by acid base transition of spinach chloroplasts. *Proceedings of the National Academy of Sciences (USA),* **55**, 170−177.

Jeffries, R. L. (1973) The ionic relations of seedlings of the halophyte, *Triglochin maritima L.* In: *Ion Transport in Plants*, ed. Anderson, W. P. Academic Press, London.

Jennings, D. H. (1963) *The absorption of solutes by plant cells.* Oliver and Boyd, Edinburgh.

Jennings, D. H. (1967) Electrical potential measurements, ion pumps and root exudation — a comment and a model explaining cation selectivity by the root. *New Phytologist* **66**, 357–369.

Jenny, H. and Overstreet, R. (1939) Cation exchange between plant roots and soil colloids. *Soil Science*, **47**, 257–272.

Jenny, H. and Grossenbacher, K. (1963) Root-soil boundary zones as seen in the electron microscope. *Soil Science Society of America Proceedings*, **27**, 273–277.

Jensen, G. (1962) Active and passive components in ion uptake processes. *Physiologia Plantarum*, **15**, 363–368.

Jeschke, W. D. (1970) Evidence for a K^+ stimulated Na^+ efflux at the plasmalemma of barley root cells. *Planta (Berlin)*, **94**, 240–245.

Jeschke, W. D. (1973) K^+ stimulated Na^+ efflux and selective transport in barley roots. In: *Ion Transport in Plants*, ed. Anderson, W. P. Academic Press, London.

Johansen, C., Edwards, D. G. and Loneragen, J. F. (1970) Potassium fluxes during potassium absorption by intact barley plants of increasing potassium content. *Plant Physiology*, **45**, 601–603.

Jyung, W. H., Wittwer, S. H. and Bukovac, M. J. (1965) Ion uptake by cells enzymically isolated from green tobacco leaves. *Plant Physiology*, **40**, 410–414.

Kahn, J. and Hansen, J. B. (1957) The effect of calcium on potassium accumulation in corn and soy bean roots. *Plant Physiology*, **32**, 312–316.

Kedem, O. (1961) Criteria of active transport. In: *Membrane Transport and Metabolism*, ed. Kleinzeller, A. and Kotyk, A. Academic Press, New York.

Knight, A. H., Crooke, W. M. and Inkson, R. H. E. (1961) Cation exchange capacities of tissues of higher and lower plants and their related uronic acid contents. *Nature (London)*, **192**, 142–143.

Koster, A. L. (1963) Changes in metabolism of isolated root systems of soy bean. *Nature (London)*, **198**, 709–710.

Kramer, P. J. (1949) *Plant soil water relationships.* McGraw Hill, New York.

Kramer, P. J. (1957) Outer space in plants. *Science*, **125**, 633–635.

Kramer, P. J. (1969) *Plant and soil water relationships: a modern synthesis.* McGraw Hill, New York.

195

Kramer, P. J. and Wiebe, H. H. (1952) Longitudinal gradients of ^{32}P absorption in roots. *Plant Physiology*, 27, 661—674.

Krasavina, M. C. and Vyskrebentseva, E. I. (1972) ATPase activity and transport of potassium and sodium in root tissues. *Fiziologya Rasteny*, 19, 978—983.

Kylin, A. and Gee, R. (1970) Adenosine triphosphatase activities in leaves of the mangrove *Avicennia nitida Jacq*. Influence of sodium to potassium ratios and salt concentrations. *Plant Physiology*, 45, 169—172.

Kylin, A. and Hylmö, B. (1957) Uptake and transport of sulphate in wheat: Active and passive components. *Physiologia Plantarum*, 10, 467—484.

Laine, T. (1934) On the absorption of electrolytes by the cut roots of plants and the chemistry of plant exudation sap. *Acta Botanica Fennica*, 16, 1—64.

Laties, G. G. (1969) Dual mechanisms of salt uptake in relation to compartmentation and long distance transport. *Annual Review of Plant Physiology*, 20, 89—116.

Laties, G. G. and Budd, K. (1964) The development of differential permeability in isolated steles of corn roots. *Proceedings of the National Academy of Sciences (USA)*, 52, 462—469.

Läuchli, A. (1967) Untersuchungen über Vorteilung und Transport von Ionen in Pflanzengeweben mit der Röntgen-Mikrosonde. I. Versuche an Vegetativen Organen von *Zea mays*. *Planta (Berlin)*, 75, 185—206.

Läuchli, A. (1972) Translocation of inorganic solutes. *Annual Review of Plant Physiology*, 23, 197—218.

Läuchli, A. and Epstein, E. (1971) Lateral transport of ions into the xylem of corn roots. I. Kinetics and energetics. *Plant Physiology*, 48, 111—117.

Läuchli, A., Kramer, D., Pitman, M. G. and Lüttge, U. (1974) Ultrastructure of xylem parenchyma cells of barley roots in relation to ion transport to the xylem. *Planta (Berlin)*, 119, 85—99.

Läuchli, A., Lüttge, U. and Pitman, M. G. (1973) Ion uptake and transport through barley seedlings: Differential effect of cycloheximide. *Zeitschrift für Naturforschung*, 28, 431—434.

Läuchli, A., Spurr, A. R. and Epstein B. (1971) Lateral transport of ions into the xylem of corn roots. II. Evaluation of a stelar pump. *Plant Physiology*, 48, 118—124.

Läuchli, A., Spurr, A. R. and Wittkop, R. W. (1970) Electron probe analysis of freeze substituted epoxy resin embedded tissue for ion transport studies in plants. *Planta (Berlin)*, 95, 341—350.

Lavey, T. L. and Barber, S. A. (1964) Movement of molybdenum in the soil and its effect on availability to the plant. *Soil Science Society of America Proceedings*, **28**, 93−97.

Lavison, J. R. (1910) Du mode de pénétration de quelques sels dans la plante vivante. *Revue Générale de Botanique*, **22** 93−97.

Leigh, R. A., Wyn Jones, R. G. and Williamson, F. A. (1973) The possible role of vesicles and ATPases in ion uptake. In: *Ion Transport in Plants*, ed. Anderson, W. P. Academic Press, London.

Leonard, R. T. and Hanson, J. B. (1972) Increased membrane-bound adenosine triphosphatase activity accompanying development of enhanced solute uptake in washed corn root tissue. *Plant Physiology*, **49**, 436−440.

Leonard, R. T. and Hodges, T. K. (1973) Characterization of plasma membrane associated adenosine triphosphate activity of oat roots. *Plant Physiology*, **52**, 6−12.

Levitt, J. (1957) The significance of 'apparent free space' (AFS) in ion absorption. *Physiologia Plantarum*, **10**, 882−888.

Lewis, D. G. and Quirk, J. P. (1967) Phosphate diffusion in soil and uptake by plants, *Plant and Soil*, **26**, 445−453.

Liberman, E. A. and Topaly, V. P. (1968) Selective transport of ions through bimolecular phospholipid membranes. *Biochimica et Biophysica Acta*, **163**, 125−136.

Loomis, W. F. and Lipman, F. (1948) Reversible inhibition of the coupling between phosphorylation and oxidation. *Journal of Biological Chemistry*, **173** 807−808.

Lopushinsky, W. and Kramer, P. J. (1961) Effect of water movement on salt movement through tomato roots. *Nature (London)*, **143**, 203−204.

Loughman, B. C. (1966) The mechanism of absorption and utilization of phosphate by barley plants in relation to subsequent transport to the shoot. *New Phytologist*, **65** 388−397.

Loughman, B. C. and Russell, R. S. (1957) The absorption and utilization of phosphate by young barley plants. IV. The initial stages of phosphate metabolism in roots. *Journal of Experimental Botany*, **8**, 280−293.

Lundegårdh, H. (1939) An electrochemical theory of salt absorption and respiration. *Nature (London)* **143** 203−204.

Lundegårdh, H. (1945) Absorption, transport and exudation of inorganic ions by the root. *Arkiv for Botanik*, A**32**, 1−139.

Lundegårdh, H. and Burström, H. (1933) Untersuchungen über die Salzaufnahme der Pflanzen. III. Quantitative Beziehungen zwischen Atmung und Anionenaufnahme. *Biochemische Zeitung*, **261**, 235−251.

Lundegårdh, H. and Burström, H. (1935) Untersuchungen über die Atmungsvorgänge in Pflanzenwurzeln. *Biochemische Zeitung*, **277**, 223–249.

Lüttge, U. and Laties, G. G. (1966) Dual mechanisms of ion absorption in relation to long distance transport in plants. *Plant Physiology*, **41**, 1531–1539.

Lüttge, U. and Laties, G. G. (1967) Absorption and long distance transport by isolated stele of maize roots in relation to the dual mechanisms of ion absorption. *Planta (Berlin)*, **74**, 173–187.

Lüttge, U. and Weigl, J. (1962) Mikroautoradiographische Untersuchungen der Aufnahme und des Transportes von $^{35}SO_4$ und ^{45}Ca in Keimwurzeln von *Zea mays L.* und *Pisum sativum*. *Planta (Berlin)*, **58**, 113–126.

Lyalin, O. O. and Ktitorova, I. H. (1969) Resting potential of the root hair of *Trianea bogotensis*. *Fiziologya Rasteny*, **16**, 261–271.

Lycklama, J. C. (1963) The absorption of ammonium and nitrate by perennial rye grass. *Acta Botanica Neerlandica*, **12**, 361–423.

Macdonald, I. R. and Ellis, R. J. (1969) Does cycloheximide inhibit protein synthesis specifically in plant tissues? *Nature (London)*, **222**, 791–792.

Machlis, L. (1944) Respiratory gradient in barley roots. *American Journal of Botany*, **31**, 281–282.

Macklon, A. E. S. (1975a) Cortical cell fluxes and transport to the stele in excised root segments of *Allium cepa L.* I. Potassium, sodium and chloride. *Planta (Berlin)*, **122**, 109–130.

Macklon, A. E. S. (1975b) Cortical cell fluxes and transport to the stele in excised root segments of *Allium cepa L.* II. Calcium. *Planta (Berlin)*, **122**, 131–141.

Macklon, A. E. S. and Higinbotham, N. (1970) Active and passive transport of potassium in cells of excised pea epicotyls. *Plant Physiology*, **45**, 133–138.

MacRobbie, E. A. C. (1962) Ionic relations of *Nitella translucens*. *Journal of General Physiology*, **45**, 861–878.

MacRobbie, E. A. C. (1969) Ion fluxes to the vacuole of *Nitella translucens*. *Journal of Experimental Botany*, **20**, 236–256.

MacRobbie, E. A. C. (1970) The active transport of ions in plant cells. *Quarterly Review of Biophysics*, **3**, 251–294.

MacRobbie, E. A. C. and Dainty, J. (1958) Ion transport in *Nitellopsis obtusa*. *Journal of General Microbiology*, **42**, 335–353.

Maizel, J. V., Benson, A. A. and Tolbert, N. E. (1956) Identification of phosphoryl choline as an important constituent of plant saps. *Plant Physiology*, **31**, 407–408.

Mason, T. G. and Maskell, E. J. (1931) Further studies on transport in the cotton plant. I. Preliminary observations on the transport of phosphorus, potassium and calcium. *Annals of Botany*, **45**, 125–173.

Meiri, A. and Anderson, W. P. (1970) Observations on the exchange of salt between the xylem and neighbouring cells in *Zea mays* primary roots. *Journal of Experimental Botany*, **21**, 908–914.

Millikan, C. R. and Hanger, R. C. (1965) Effects of chelation and of certain cations on the mobility of foliar applied [45]Ca in stock, broad bean, peas and subterranean clover. *Australian Journal of Biological Sciences*, **18**, 211–226.

Mitchell, P. (1961) Coupling of phosphorylation to electron and hydrogen transfer by a chemi-osmotic type mechanism. *Nature (London)*, **191**, 144–145.

Mitchell, P. (1966) Chemi-osmotic coupling in oxidative and photosynthetic phosphorylation. *Biological Reviews*, **44**, 445–502.

Mitchell, P. (1970) Membranes of cells and organelles: Morphology, transport and metabolism. *Symposium of the Society for General Microbiology*, **20**, 121–166.

Mitchell, P. (1973) Performance and conservation of osmotic work by proton-coupled solute porter systems. *Journal of Bioenergetics*, **4**, 63–91.

Mitchell, P. and Moyle, J. (1965) Stoichiometry of proton translocation through the respiratory chain and adenosine triphosphatase systems of rat liver mitochondria. *Nature (London)*, **208**, 147–151.

Montasir, A. H., Sharoubeem, H. H. and Sidrak, G. H. (1966) Partial substitution of sodium for potassium in water cultures. *Plant and Soil*, **25**, 181–193.

Morrison, T. M. (1965) Xylem sap composition in woody plants. *Nature (London)*, **205**, 1027.

Nassery, H. and Baker, D. A. (1972) Extrusion of sodium ions by barley roots. I. Characteristics of the extrusion mechanism. *Annals of Botany*, **36**, 881–887.

Nemček, O., Sigler, K., and Kleinzeller, A. (1966) Ion transport in the pitcher of *Nepenthes henryana*. *Biochimica et Biophysica Acta*, **126**, 73–80.

Nicholson, T. H. (1967) Vesicular-arbuscular mycorrhiza – a universal plant symbiosis. *Science Progress (London)*, **55**, 561–581.

Nissen, P. (1971) Uptake of sulphate by roots and leaf slices of barley: Mediated by single, multiphasic mechanisms. *Physiologia Plantarum*, **24**, 315–324.

Nissen, P. (1973) Multiphasic uptake in plants. II. Mineral cations, chloride and boric acid. *Physiologia Plantarum,* 29, 298–254.

Nobel, P. S. (1970) *Plant cell physiology.* Freeman, San Francisco.

Nye, P. H. (1966) The effect of nutrient intensity and buffering power of a soil, and the absorbing power, size and root hairs of a root, on nutrient absorption by diffusion. *Plant and Soil,* 25, 81–105.

Nye, P. H. and Marriott, F. H. C. (1968) The importance of mass flow in the uptake of ions by roots from soil. *Transactions of the 9th Congress of the International Soil Science Society,* 1, 127–134.

Olsen, C. (1950) The significance of concentration for the rate of ion absorption by higher plants in water culture. *Physiologia Plantarum,* 3, 152–164.

Olsen, C. (1953) The significance of concentration for the rate of ion absorption by higher plants in water culture. IV. The influence of hydrogen ion concentration. *Physiologia Plantarum,* 6, 848–855.

Olsen, R. A. and Peech, M. (1960) The significance of the suspension effect in the uptake of cations by plants from soil-water systems. *Soil Society of America Proceedings,* 24, 257–260.

Osterhout, W. J. V. (1936) The absorption of electrolytes in large plant cells. *The Botanical Review,* 2, 283–315.

Pallaghy, C. K., Lüttge, U. and von Willert, K. (1970) Cytoplasmic compartmentation and parallel pathways of ion uptake in plant root cells. *Zeitschrift fur Pflanzenphysiologie,* 62, 51–57.

Pallaghy, C. K. and Scott, B. I. H. (1969) The electrochemical state of cells of broad bean roots. II: Potassium kinetics in excised root tissue. *Australian Journal of Biological Sciences,* 22, 585–600.

Pardee, A. B. (1966) Membrane transport proteins. *Science,* 162, 632–637.

Passioura, J. B. (1963) A mathematical model for the uptake of ions from the soil solution. *Plant and Soil,* 18, 225–238.

Pate, J. S. (1965) Roots as organs of assimilation of sulphate. *Science,* 149, 547–548.

Pearson, G. A. (1967) Absorption and translocation of sodium in beans and cotton. *Plant Physiology,* 42, 1171–1175.

Persson, L. (1969) Labile-bound sulphate in wheat roots: Localization, nature and possible connection to the active absorption mechanism. *Physiologia Plantarum,* 22, 959–976.

Pettersson, S. (1960) Ion absorption in young sunflower plants. I. Uptake and transport mechanisms for sulphate. *Physiologia Plantarum,* 13, 133–147.

Pettersson, S. (1966) Active and passive components of sulphate in sunflower plants. *Physiologia Plantarum,* 19, 459–492.

Pettersson, S. (1971) A labile-bound component of phosphate in the free space of sunflower plant roots. *Physiologia Plantarum*, 24, 485–490.

Pierce, W. S. and Higinbotham, N. (1970) Compartments and fluxes of K^+, Na^+ and Cl^- in *Avena* coleoptile cells. *Plant Physiology*, 46, 666–673.

Pitman, M. G. (1963) The determination of the salt relations of the cytoplasmic phase in cells of beetroot tissue. *Australian Journal of Biological Sciences*, 16, 647–668.

Pitman, M. G. (1965) Transpiration and the selective uptake of potassium by barley seedlings (*Hordeum vulgare* cv Bolivia). *Australian Journal of Biological Sciences*, 18, 987–988.

Pitman, M. G. (1966) Uptake of potassium and sodium by seedlings of *Sinapsis alba*. *Australian Journal of Biological Sciences*, 19, 257–269.

Pitman, M. G. (1970) Active H^+ efflux from cells of low salt barley roots during salt accumulation. *Plant Physiology*, 45, 787–790.

Pitman, M. G. (1971) Uptake and transport of ions in barley seedlings. I. Estimation of chloride fluxes in cells of excised barley roots. *Australian Journal of Biological Sciences*, 24, 407–421.

Pitman, M. G. (1972) Uptake and transport of ions in barley seedlings. II. Evidence for two active stages in transport to the shoot. *Australian Journal of Biological Sciences*, 25, 243–257.

Pitman, M. G., Courtice, A. C. and Lee, B. (1968) Comparison of potassium and sodium uptake by barley roots at high and low salt status. *Australian Journal of Biological Sciences*, 21, 871–881.

Pitman, M. G. and Cram, W. J. (1973) Regulation of inorganic ion transport in plants. In: *Ion Transport in Plants*, ed., Anderson, W. P. Academic Press, London.

Pitman, M. G., Mowat, J. and Nair, H. (1971) Interaction of processes for accumulation of salt and sugar in barley plants. *Australian Journal of Biological Sciences*, 24, 619–631.

Pitman, M. G. and Saddler, H. D. W. (1967) Active sodium and potassium transport in cells of barley roots. *Proceedings of the National Academy of Sciences (USA)*, 57, 44–49.

Polle, E. O. and Jenny, H. (1971) Boundary layer effects in ion absorption by roots and storage organs of plants. *Physiologia Plantarum*, 25, 219–224.

Poux, N. (1967) Localisation d'activités enzymatiques dans les cellules du méristème radiculaire de *Cucumis sativus* L. *Journal de Microscopie*, 6, 1043–1058.

Power, J. B. and Cocking, E. C. (1970) Isolation of leaf protoplasts;

macromolecule uptake and growth substance response. *Journal of Experimental Botany*, **21**, 64—70.

Pressman, B. C. (1968) Ionophorous antibiotics as models for biological transport. *Federation Proceedings of the American Societies for Experimental Biology*, **27**, 1283—1288.

Prevot, P. and Steward, F. C. (1936) Salient features of the root system relative to the problem of salt absorption. *Plant Physiology*, **11**, 509—534.

Randall, J. (1974) The Electron microscopy and composition of biological membranes and envelopes. *Philosophical Transactions of the Royal Society of London*, **B 268**, 1—159.

Rao, L. C., Krishnamurthy, T. N. and Rao, J. T. (1967) Cation exchange capacity of roots and yield potential in sugar cane. *Plant and Soil*, **27**, 314—318.

Ratner, A. and Jacoby, B. (1973) Non specificity of salt effects on Mg dependent ATPase from grass roots. *Journal of Experimental Botany*, **24**, 231—238.

Raven, J. A. (1971) Ouabain-insensitive K influx in *Hydrodictyon africanum. Planta (Berlin)*, **97**, 28—38.

Raven, J. A. and Smith, F. A. (1973) The regulation of intracellular pH as a fundamental biological process. In: *Ion Transport in Plants*, ed., Anderson, W. P. Academic Press, London.

Reid, R. A., Moyle, J. and Mitchell, P. (1966) Synthesis of adenosine triphosphate by a proton motive force in rat liver mitochondria. *Nature (London)*, **212**, 257—258.

Robards, A. W. (1971) The ultrastructure of plasmodesmata. *Protoplasma*, **72**, 315—323.

Robards, A. W. and Robb, M. E. (1972) Uptake and binding of uranyl ions by barley roots. *Science*, **178**, 980—982.

Robertson, R. N. (1968) Protons, electrons, phosphorylation and active transport. University Press, Cambridge.

Robertson, R. N. and Turner, J. S. (1945) Studies in the metabolism of plant cells. III. The effects of cyanide on the accumulation of potassium chloride and on respiration: The nature of salt respiration. *Australian Journal of Experimental Biology and Medical Science*, **23**, 63—73.

Robertson, R. N., Wilkins, M. J. and Weeks, D. C. (1951) Studies in the metabolism of plant cells. IX. The effects of 2,4-dinitrophenol on salt accumulation and salt respiration. *Australian Journal of Scientific Research*, B **4**, 248—264.

Rovira, A. D. and Bowen, G. D. (1968) Anion uptake by the apical region of seminal wheat roots. *Nature (London)*, **218**, 685—686.

Rowell, D. L., Martin, M. W. and Nye, P. H. (1967) The measurement and mechanism of ion diffusion in soils. *Journal of Soil Science*, **18**, 204–222.

Rungie, J. M. and Wiskitch, J. T. Salt stimulated adenosine triphosphatases from smooth microsomes of turnip. *Plant Physiology*, **51**, 1064–1068.

Russell, R. S. and Shorrocks, V. M. (1959) The relationship between transpiration and the absorption of inorganic ions by intact plants. *Journal of Experimental Botany*, **10**, 301–316.

Sabinin, D. A. (1925) On the root system as an osmotic apparatus. *Bulletin de L'Institut des Recherches Biologiques, Université de Perm*, **4**, Supplement 2, 1–136.

Scott, B. I. H., Gulline, H. and Pallaghy, C. K. (1968) The electrochemical state of cells of broad bean root. I. Investigations of elongating roots of young seedlings. *Australian Journal of Biological Sciences*, **21**, 185–200.

Scott, F. M. (1949) Plasmodesmata in xylem vessels. *Botanical Gazette*, **110**, 492–495.

Scott, L. I. and Priestley, J. H. (1928) The root as an absorbing organ. I. A reconsideration of the entry of water and salt in the absorbing region. *New Phytologist*, **27**, 125–140.

Shepherd, U. H. and Bowling, D. J. F. (1973) Active accumulation of sodium by roots of five aquatic species. *New Phytologist*, **72**, 1075–1080.

Shone, M. G. T. (1968) Electrochemical relations in the transfer of ions to the xylem sap of maize roots. *Journal of Experimental Botany*, **19**, 468–485.

Shone, M.G. T. (1969) Origins of the electrical potential difference between the xylem sap of maize roots and the external solution. *Journal of Experimental Botany*, **20**, 698–716.

Shone, M. G. T., Clarkson, D. T., Sanderson, J. and Wood, A. V. (1973) A comparison of the uptake and translocation of some organic molecules and ions in higher plants. In: *Ion Transport in Plants*, ed. Anderson, W. P. Academic Press, London.

Sinyukhin, A. M. and Vyskrebentseva, E. I. (1967) The influence of potassium ions in the resting potential of the root cells of *Cucurbita pepo*. *Soviet Plant Physiology*, **14**, 553–557.

Smith, R. C. and Epstein, E. (1964) Ion absorption by shoot tissue: Technique and first findings with excised leaf tissue of corn. *Plant Physiology*, **39**, 338–341.

Spanswick, R. M. (1972) Evidence for an electrogenic ion pump in *Nitella translucens*. I. The effects of pH, K^+, Na^+, light and temperature on the membrane potential and resistance. *Biochimica et Biophysica Acta*, **288**, 73—89.

Spanswick, R. M. (1973) Electrogenesis in photosynthetic tissues. In: *Ion Transport in Plants*, ed. Anderson, W. P. Academic Press, London.

Spanswick, R. M. (1974) Evidence for an electrogenic pump in *Nitella translucens*. II. Control of the light stimulated component of the membrane potential. *Biochimica et Biophysica Acta*, **332**, 387—398.

Spanswick, R. M. and Williams, E. J. (1964) Electrical potentials and Na^+, K^+ and Cl^-, concentrations in the vacuole and cytoplasm of *Nitella translucens*. *Journal of Experimental Botany*, **15**, 193—200.

Steward, F. C. (1932) The absorption and accumulation of solutes by living plant cells. I. Experimental conditions which determine salt absorption by storage tissue. *Protoplasma*,**15**, 29—58.

Steward, F. C. (1933) The absorption and accumulation of solutes by living plant cells. V. Observations on the effects of time, oxygen and salt concentration upon absorption and respiration by storage tissue. *Protoplasma*, **18**, 208—242.

Steward, F. C. (1943) The effect of ringing and transpiration on mineral uptake. *Annals of Botany* (*New Series*), **7**, 89—92.

Steward, F. C. and Berry, W. E. (1934) The absorption and accumulation of solutes by living plant cells. VII The time factor in the respiration and salt absorption of Jerusalem artichoke tissue (*Helianthus tuberosus*) with observations on ionic interchange. *Journal of Experimental Biology*, **11**, 103—119.

Steward, F. C., Berry, W. E. and Broyer, T. C. (1936) The absorption and accumulation of solutes by living plant cells. VIII. The effect of oxygen upon respiration and salt accumulation. *Annals of Botany* (*London*), **50**, 345—366.

Steward, F. C., Prevot, P. and Harrison, J. A. (1942) Absorption and accumulation of rubidium bromide by barley roots. Localisation in the root of cation accumulation and of transfer to the shoot. *Plant Physiology*, **17**, 411—421.

Steward, F. C. and Sutcliffe, J. F. (1959) Plants in relation to inorganic salts. In: *Plant Physiology — a Treatise*, II, ed. Steward, F. C. Academic Press, London.

Stout, P. R. and Hoagland, D. R. (1939) Upward and lateral movement of salt in certain plants as indicated by radioactive isotopes of potassium, sodium and phosphorus absorbed by roots. *American Journal of Botany*, **26**, 320—324.

Sutcliffe, J. F. (1956) The selective absorption of alkali cations by storage tissues and intact barley plants. Potash Symposium, 1956, pp. 1—11, International Potash Institute, Berne.

Sutcliffe, J. F. (1962) *Mineral salt absorption in plants.* Pergamon Press, Oxford.

Tanada, T. (1955) The affect of UV radiation and calcium and their interaction on salt absorption by excised mung bean roots. *Plant Physiology,* **30**, 221—225.

Tanton, T. W. and Crowdy, S. H. (1972) Water pathways in higher plants. II. Water pathways in roots. *Journal of Experimental Botany,* **23**, 600—618.

Thellier, M. (1970) An electrokinetic interpretation of the functioning of biological systems and its application to the study of mineral salts absorption. *Annals of Botany,* **34**, 983—1009.

Thellier, M., Thoiron, B. and Thoiron, A. (1971) Electrokinetic formulation of overall kinetics of *in vivo* processes. *Physiologie Végétale,* **9**, 65—82.

Tiffin, L. O. (1967) Translocation of manganese, iron, cobalt and zinc in tomato. *Plant Physiology,* **42**, 1427—1432.

Tinker, P. B. (1969) The transport of ions in the soil around plant roots. In: *Ecological Aspects of Mineral Nutrition of Plants,* ed. Rorison, I. H. Blackwell, Oxford.

Torii, K. and Laties, G. G. (1966) Dual mechanisms of ion uptake in relation to vacuolation in corn roots.*Plant Physiology,* **41**, 863—870.

Tromp, J. (1962) Interactions in the absorption of ammonium, potassium and sodium ions by wheat roots. *Acta Botanica Neerlandica,* **11**, 147—192.

Tukey, H. B. and Mecklenburg, R. A. (1964) Leaching of metabolites from foliage and subsequent reabsorption and redistribution of the leachate in plants. *American Journal of Botany,* **51**, 737—742.

Tyree, M. T. (1970) The symplasm concept. A general theory of symplastic transport according to the thermodynamics of irreversible processes. *Journal of Theoretical Biology,* **26**, 181—214.

Ulrich, A. (1941) Metabolism of non-volatile organic acids in excised barley roots as related to cation-anion balance during salt accumulation. *American Journal of Botany,* **28**, 526—537.

Ussing, H. H. (1949) The distinction by means of tracers between active transport and diffusion. *Acta Physiologia Scandinavica,* **19**, 43—56.

Vaadia, Y. (1960) Autonomic diurnal fluctuations in rate of exudation and root pressure of decapitated sunflower plants. *Physiologia Plantarum,* **13**, 707—717.

Van Andel, O. M. (1953) The influence of salts on the exudation of tomato plants. *Acta Botanica Neerlandica*, **2**, 445–521.

Van Fleet, D. S. (1961) Histochemistry and function of the endodermis. *Botanical Review*, **27**, 165–220.

Van den Honert, T. H. and Hooymans, J. J. M. (1955) On the absorption of nitrate by maize in water culture. *Acta Botanica Neerlandica*, **4**, 376–384.

Van den Honert, T. H., Hooymans, J. J. M. and Volkers, W. S. (1955) Experiments on the relation between water absorption and mineral uptake by plant roots. *Acta Botanica Neerlandica*, **4**, 139–155.

Van Overbeek, J. (1942) Water uptake by excised root systems of the tomato due to non osmotic forces. *American Journal of Botany*, **29**, 677–683.

Viets, F. G. (1944) Calcium and other polyvalent cations as accelerators of ion accumulation by excised barley roots. *Plant Physiology*, **19**, 466–480.

Walker, J. M. and Barber, S. A. (1962) Absorption of potassium and rubidium from the soil by corn roots. *Plant and Soil*, **17**, 243–259.

Wallace, A. and North, C. P. (1953) Lime induced chlorosis. *California Agriculture*, **7**, 10.

Weigl, J. and Lüttge, U. (1962) Mikroautoradiographische Untersuchungen der Aufnahme von $^{35}SO_4$ durch Wurzeln von *Zea mays L.* Die Funktion der primären Endodermis. *Planta (Berlin)*, **59**, 15–28.

Weissmann, G. S. (1964) Effect of ammonium and nitrate nutrition on protein level and exudate composition. *Plant Physiology*, **39**, 947–952.

Welch, R. M. and Epstein, E. (1968) The dual mechanisms of alkali cation absorption by plant cells: Their parallel operation across the plasmalemma. *Proceedings of the National Academy of Science (USA)*, **61**, 447–453.

Welch, R. M. and Epstein, E. (1969) The plasmalemma: Seat of the type 2 mechanisms of ion absorption. *Plant Physiology*, **44**, 301–304.

Whittam, R. and Wheeler, K. P. (1970) Transport across cell membranes. *Annual Review of Physiology*, **32**, 21–60.

Willbrandt, W. (1954) Secretion and transport of non-electrolytes. *Symposium of the Society for Experimental Biology*, **8**, 136–162.

Williams, D. E. and Coleman, N. T. (1950) Cation exchange properties of plant root surfaces. *Plant and Soil*, **2**, 243–256.

Willison, J. H. M., Grout, B. W. W. and Cocking, E. C. (1971) A mechanism for the pinocytosis of latex spheres by tomato fruit protoplasts. *Bioenergetics*, 2, 371–382.

Woodford, E. K. and Gregory, F. G. (1948) Preliminary results obtained with an apparatus for the study of salt uptake and root respiration of whole plants. *Annals of Botany, (London) New Series*, 12, 335–370.

Yoder, O. C. and Scheffer, R. P. (1973) Effects of *Helminthosporium carbonum* toxin on absorption of solutes by corn roots. *Plant Physiology*, 52, 518–523.

Yu, G. H. and Kramer, P. J. (1967) Radial salt transport in corn roots. *Plant Physiology*, 42, 985–990.

Yu, G. H. and Kramer, P. J. (1969) Radial transport of ions in roots. *Plant Physiology*, 44, 1095–1100.

Index